Lecture Notes
in Control and Information Sciences 191

Editor: M. Thoma

Vincent Blondel

Simultaneous Stabilization of Linear Systems

Springer-Verlag London Ltd.

Author

Vincent Blondel, PhD
Department of Mathematics/Division of Systems and Optimization, Royal Institute
of Technology (KTH), S-100 44 Stockholm, Sweden

ISBN 978-3-540-19862-8   ISBN 978-3-540-39321-4 (eBook)
DOI 10.1007/978-3-540-39321-4

Typesetting: Camera ready by author

69/3830-543210 Printed on acid-free paper

To Gérard Valenduc,
my secondary school teacher in physics.
To Michel Gevers,
my Ph.D. thesis supervisor.
And to all those who devote part of their life transmiting
their enthousiasm for the scientific adventure.

Preface

The research reported in this monograph (mainly the last three chapters) was performed while I was preparing a Ph.D. thesis in Applied Mathematics at the Catholic University of Louvain (Louvain-la-Neuve, Belgium) and during extended visits to the Australian National University (Canberra, Australia), Imperial College (London, U.K.) and Oxford University (Oxford, U.K.).

I am pleased to acknowledge F. Callier, G. Campion, S. Dasgupta, M. Gevers, H. Kwakernaak, J. Mawhin and J. Meinguet for their evaluation of an earlier version of this work and I am particularely thankful to M. Gevers and H. Kwakernaak for their scientific but also professional help.

Some of the ideas expressed in the monograph resulted from exchanges on simultaneous stabilization with several researchers whom I gratefully acknowledge here: B. Anderson (Canberra, Australia), S. Dasgupta (Iowa, U.S.A.), P. Delsarte (Louvain-la-Neuve, Belgium), M. Fu (Newcastle, Australia), M. Green (Canberra, Australia), R. Rupp (Karlsruhe, Germany), E. Sontag (Rutgers, U.-S.A.), M. Vidyasagar (Bangalore, India) and K. Wei (DLR, Germany).

Professor M. Thoma (Hannover, Germany), the Editor of this series, has been particularly encouraging and enthousiastic concerning this volume. He has been of crucial support.

My last acknowledgments go, for their sacrifice, to all English speaking persons. By instituting their mother tongue as the 'international language of science' they have transformed it into the most tortured language of all times[1]. This monograph, written in the purest 'broken English style' is no exception and will most certainly add to their rising agony.

Roûmont-sur-Ourthe,

April 1993.

---

[1]This could not have been the case with French: "President de Gaulle once insisted that French should speak French at international conferences. This was countered by the English participants who said that they would speak French too!"

Prologue

"What is the use of a book", thought Alice,
"without pictures or conversation".

L. Carroll, Alice's adventures in wonderland.

Catholic University of Louvain, Louvain-la-Neuve, Belgium, 1989.

V.B. *Heu... Soura, I don't know much about systems and control theory... may sound a bit silly but... when is a system stabilizable?*

S.D. *Depends. What kind of system and what kind of controller?*

V.B. *The simplest I can think of... finite systems and controllers that are linear time invariant,... scalar and... that's the simplest case I can think of.*

S.D. *With these assumptions a system is always stabilizable provided no unstable pole-zero cancellations occur.*

V.B *Very well.*

And what are the controllers that stabilize a given system?

S.D. Have a look at Vidyasagar's book or at one of the original papers of Youla or Kucera. They provide a constructive procedure for the set of all the stabilizing controllers of a given system.

V.B. Good...

Ahem... sorry but for two? When are two systems simultaneously stabilizable?

S.D. Again, that question was fully solved by Vidyasagar and Viswanadhan and by Saeks and Murray some ten years ago. Roughly speaking, if $p_1(s)$ and $p_2(s)$ are the transfer functions of the two systems, then these two systems are simultaneously stabilizable if and only if the system whose transfer function is defined by $p(s) = p_1(s) - p_2(s)$ is stabilizable by a stable controller.

V.B. Allright, thanks.

[...]

V.B. Hum... sorry to bother you again, was thinking and... when is a system stabilizable by a stable controller?

S.D If and only if the system has an even number of unstable zeros between each pair of unstable poles on the real axis.

V.B. Woahh... that's a surprising condition.

S.D. Yeap... and an elegant one.

V.B. ...

S.D. Want a hard question?

V.B. *Please do.*

S.D. *When are three systems simultaneously stabilizable?*

V.B. *Sorry but I don't know. It doesn't look too hard... what is the solution?*

S.D. *I don't know either and in fact nobody knows... the question has been open for more than ten years.*

V.B. *Such a simply stated question? Are you joking?*

S.D. *I'm serious.*

V.B. *Well then...*

[...] see the Epilogue (page 145).

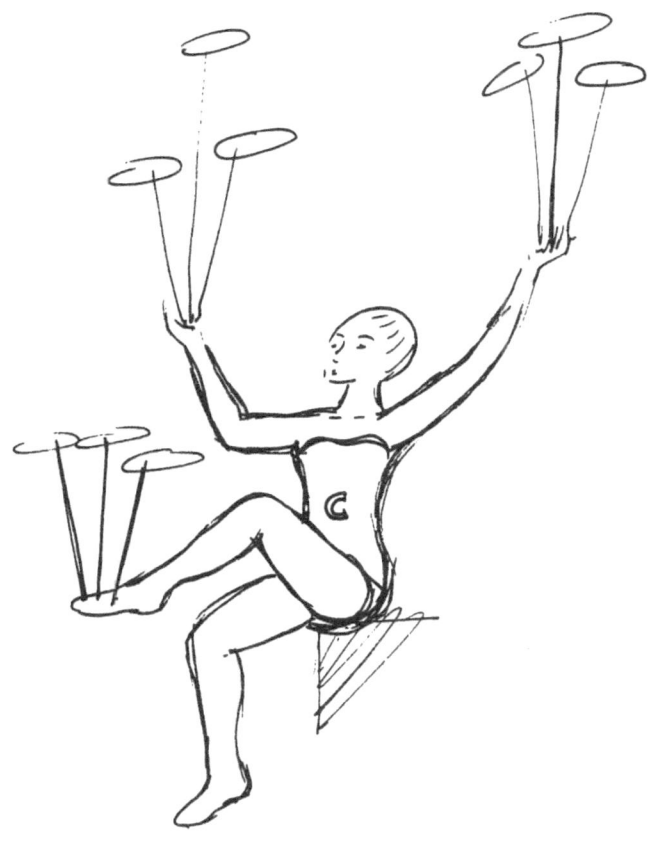

Simultaneous stabilization, © Isabelle de Failly, 1993.

Notations and conventions

Square brackets [..] are used for references.

Real and complex variables

$\pi = 3.1415$ and $e = 2.7172$.

$\bar{s}$ is the complex conjugate of $s \in C$.

$\Re(s)$ is the real part of $s \in C$.

$\Im(s)$ is the imaginary part of $s \in C$.

Depending on the context we use the symbols s, λ or z to denote complex variables.

Subsets of the extended complex plane

N, Z, Q, R and C are the sets of natural, integer, rational, real and complex numbers.

∞ is the point at infinity.

$C_\infty = C \cup \{\infty\}$ is the extended complex plane. This set is endowed with the usual Riemann sphere topology and is sometimes denoted by $\hat{C}$ in the literature.

$C_+ = \{s \in C : \Re(s) \geq 0\}$ is the right half plane.

$C_{+\infty}$ is the extended right half plane. It is a closed subset of C_∞.

R_∞, R_+ and $R_{+\infty}$ are defined similarly to C_∞, C_+ and $C_{+\infty}$.

$D = \{z \in C : |z| < 1\}$ is the open unit disc.

$\overline{D} = \{z \in C : |z| \leq 1\}$ is the closed unit disc.

$\partial D = \{z \in \mathbf{C} : |z| = 1\}$ is the boundary of D, i.e. the unit circle.

$B(a, \rho) = \{s \in \mathbf{C} : |s - a| < \rho\}$ is the open ball centered at a and of radius ρ.

$\overline{B}(a, \rho) = \{s \in \mathbf{C} : |s - a| \leq \rho\}$ is the closed ball centered at a and of radius ρ.

Λ is used to denote any subset of $\mathbf{C}_\infty$ such that $\Lambda \cap \mathbf{R}_\infty \neq \mathbf{R}_\infty$.

Sets of functions

$\mathbf{R}[s]$ is the set of real polynomials in the variable s.

$\mathbf{R}(s)$ is the set of real rational functions in the variable s.

$\mathbf{Q}(\beta)$ is the set of rational functions in the variable β and with rational coefficients.

$\mathbf{R}'(s)$ is the set of real rational functions together with ∞.

$\mathbf{C}[s]$ and $\mathbf{C}(s)$ are defined similarly as $\mathbf{R}[s]$ and $\mathbf{R}(s)$.

$S(\Lambda)$ is the set of real rational functions with no poles in $\Lambda \subset \mathbf{C}_\infty$. These functions are called Λ-stable.

$U(\Lambda)$ is the set of real rational functions with no poles neither zeros in $\Lambda \subset \mathbf{C}_\infty$. These functions are called Λ-bistable.

S and U are used as shorthands for $S(\mathbf{C}_{+\infty})$ and $U(\mathbf{C}_{+\infty})$.

$H(D)$ is the set of analytic functions on D.

H_∞ is the set of functions that are analytic in the interior of $\mathbf{C}_{+\infty}$ and bounded on $\mathbf{C}_{+\infty}$.

$A(\overline{D})$ is the set of analytic functions on D that are also continuous on $\overline{D}$. This set is called the *disc algebra*.

Miscellaneous

$A^{m \times n}$ is the $m \times n$ matrix with entries in A.

$\Gamma(.)$ is the gamma function, $\Gamma(n + 1) = (n + 1).\Gamma(n)$. The symbol $!$ is used to denote the function Γ for $n \in \mathbf{N}$.

I_n is the $n \times n$ identity matrix.

ln is the logarithm in base e.

q is an arbitrary element of $\mathbf{R}(s)$.

r is an arbitrary element of S.

R is an arbitrary commutative ring with identity.

Wp is the set of winding transforms of p.

$W_{s_0}^{s_1}p$ is the winding number of p between s_0 and s_1.

When considering a function defined on a set P and taking values in a set V we call the elements of P *points* and the elements of V *values*. The function f maps points to values.

Contents

1 Introduction   **1**

  1.1 Background   1

  1.2 Organization 11

2 Stabilization   **15**

  2.1 Introduction 15

  2.2 Definitions 16

  2.3 Algebraic framework 22

      2.3.1 Introduction 22

      2.3.2 Ring of stable rational functions 22

      2.3.3 Ring concepts and stabilization 25

  2.4 Geometrical framework 29

      2.4.1 Introduction 29

      2.4.2 Intersection and avoidance 30

      2.4.3 Avoidance and stabilization 36

  2.5 General setting 39

  2.6 Summary and bibliography 43

3 Youla-Kucera parametrization   **46**

  3.1 Introduction 46

  3.2 The parametrization 47

  3.3 Finiteness and properness 53

3.4 Equivalences 59

    3.4.1 Simultaneous and strong stabilization 60

    3.4.2 Simultaneous and bistable stabilization . . . 62

3.5 Summary and bibliography 66

4 Necessary conditions: interlacement   68

4.1 Introduction 68

4.2 Two systems and strong stabilization 71

4.3 Non intersecting systems 75

    4.3.1 Introduction 75

    4.3.2 The 3-interlacing condition 76

    4.3.3 The even interlacing condition 82

    4.3.4 The k-interlacing condition 84

4.4 Intersecting systems 88

    4.4.1 Introduction 88

    4.4.2 Winding transform 89

    4.4.3 Properties of winding transforms 91

    4.4.4 Intersecting systems 96

4.5 Summary and bibliography 100

5 Sufficient conditions: special cases   102

5.1 Introduction 102

5.2 Two systems and strong stabilization 104

5.3 Sufficiency of interlacement conditions 112

    5.3.1 Introduction 112

    5.3.2 Four systems 113

    5.3.3 Three systems 116

    5.3.4 Bistable stabilization 118

5.4 Sufficient conditions for more than two systems . . 124

5.5 Sufficient conditions from H_∞ 127

5.6   Summary and bibliography   129

6   Necessary and sufficient conditions: rational decid-
    ability   132
    6.1   Introduction   132
    6.2   Rational decidability and algebraic numbers   134
    6.3   Simultaneous stabilization of 3 or more systems . .   139
    6.4   Summary and bibliography   144

Epilogue   145

A   Rings and Algebras   151

B   Analytic functions   157

C   Range of analytic functions   160
    C.1   Filled ball results   161
    C.2   Empty ball results   166

Bibliography   169

Index   183

Chapter 1

Introduction

A mathematical theory is not to be considered complete until you have made it so clear that you can explain it to the first man in the street.

D. Hilbert, Archiv der Mathematik und Physik, 1901.

1.1 Background

Robust control has emerged this last decade as an important issue in control design. The idea of robustness requirement is easily grasped, it expresses the legitimate wish of the control system designer to keep good controller performance in the face of system uncertainties or modifications.

Assume that we possess a model of a physical system. Based on this model we design a stabilizing controller that perhaps also achieves additional performance objectives. Unfortunately, the model is only an approximation of the real physical system and the idea of robust control design is to implement a controller that achieves reasonably good performance for the model and also for a neighbourhood of models in which the physical system is likely to be. This controller is then called robust.

There exist many sensible ways to describe the uncertainties associated to a model. To each of these descriptions corresponds a particular robust control technique. Horowitz's method [66] for example is an engineering minded technique that essentially consists into a set of practical geometrical rules. This technique can be successfuly applied to many real life situations but it also leaves several important theoretical questions unanswered, in particular when the technique fails to find a stabilizing controller.

More recent is the approach that originated from a remarkable result due to the Russian mathematician Vladimir Kharitonov. Inspired by a footnote of Gantmacher's book (Gantmacher [48]), Kharitonov wrote one of the most widely quoted papers in control theory[1]. A family of real polynomials whose coefficients lie in closed intervals is a family of stable polynomials (i.e. polynomials with no roots in the right half plane) if and only if four distinguished members of the family are stable (for a clear and succinct proof of Kharitonov's theorem, see Dasgupta [30]). Kharitonov's original paper has had a large and varied descendance since its 1976 birthdate. The descendants are usually of the form:

---

[1] Let's add one occurence: it is reference [77].

assume that $\mathcal{P}$ describes a family of polynomials whose coefficients depend in some particular way (linear, affine, multilinear, polynomial, non linear, etc) on unknown parameters that live in some subsets of the complex plane. Under what condition do all the polynomials of $\mathcal{P}$ have no roots in a given subset Ω of the complex plane?

The relevance of these sorts of results for robust control of linear systems is immediate. If a linear system has unknown coefficients that lie within some bounds, then the stability of the systems in feedback with a controller can be checked by using Kharitonov's theorem on the denominators of the closed loop transfer functions. This technique has, however, the major disadvantage of being by essence an a-posteriori method. It allows one to check the robustness of a controller once this controller is chosen but it does not really help to chose a controller. It is therefore not a design procedure.

More effective with respect to design strategy is the brainchild of Gordon Zames and Bruce Francis [42,43,44]. Their H_∞ approach to robust control design is so far the most successful and satisfactory answer given to the original practical problem: a formulation of the robust control objective as one of stabilizing all systems lying in a ball centered on a nominal model allows one to express stabilization conditions in an elegant, and solvable, mathematical way. The area of H_∞ control has given rise to books, conferences, hundreds of papers and many fruitful applications. These techniques are now included as standard routines in control software packages.

These three theories all deal with infinitely many systems. They are of the form:

> Let $\mathcal{P}$ be a particular (uncountable) infinite family of systems. Try to find a controller C that is stabilizing for each $P \in \mathcal{P}$.

Simultaneous stabilization can be seen in this context of robust control but is rather different with respect to the number of systems involved. The central question in simultaneous stabilization is:

> Let P_1, P_2,..., P_k be k systems. Under what condition is it possible to find a controller C that is stabilizing for each P_i $(i = 1, ..., k)$?

Thus the crucial difference with the above mentioned techniques is that only finitely many systems are considered: the uncertainty on the system is described in a finite way.

Finite descriptions of uncertainty appear naturally in practical situations. For example, the k systems may represent a nominal system and many of its failed modes (Emre [39], Saeks and Murray [102]), a sytem that has several operating points (such as an airplane that takes-off, flies and (hopefully) lands[2] (Ackermann [2])), or a multivariable system with possible loss of sensors or actuator failures (Alos [3]). We refer the reader to the monograph of Ackermann [2] for many more illustrative examples of simultaneous stabilization.

---

[2] We quote from a personal communication of Professor M. Vidyasagar "My past experience indicates that industrial control system designers (especially aerospace engineers) are much more excited by the simultaneous stabilization problem [than by classic robust control]".

Another practical motivation for looking at simultaneous stabilization is that the question is closely connected to strong stabilization (stabilization with a stable controller) and bistable stabilization (stabilization with a stable and minimum phase controller). Real motivations exist for these two design strategies since control designers are usually understandably reluctant to use unstable controllers even though unstable controllers may sometimes perform better than stable ones.

Simultaneous stabilization of linear sytems may appear to be included in the classical robust control frameworks of Horowitz, Kharitonov or Francis-Zames. Indeed, by including the systems P_1, P_2, ..., P_k in a larger set of systems $\mathcal{P}$, the stabilizability of the set $\mathcal{P}$ then implies that of P_1, P_2, ..., P_k and so sufficient conditions for stabilization of $\mathcal{P}$ are also sufficient for simultaneous stabilizability of P_1, P_2, ..., P_k. However, such an approach leads to deceptively conservative results. It often happens that although the systems P_1, P_2, ..., P_k are simultaneously stabilizable, the set $\mathcal{P}$ including them all is not. Moreover, including the different systems P_1, P_2, ..., P_k in some continuum description is equivalent to loosing the power and essence of the idea of simultaneous stabilization for which the systems P_1, P_2, ..., P_k are not supposed to maintain any relations among them.

In fact, and however strange it may appear at first, testing whether a finite number of systems are simultaneously stabilizable is a question that is considerably harder to solve than the one where a continuum of systems is considered. To support this statement, assume for example that we consider the family of systems described by $\mathcal{P}$

$= \{\lambda p(s) : \lambda \in [0,1]\}$ where $p(s)$ is a finite, proper, linear time invariant, scalar system of degree n $(p(s) \in \mathbf{R}(s))$. It is known (see for example Ghosh [55]) that if a stabilizing controller exists for $\mathcal{P}$ then the controller may be chosen with degree strictly less than $3n - 1$. Therefore, in order to check if a stabilizing controller exists for $\mathcal{P}$ it suffices to inject a parametrization of all the controllers of degree $3n - 2$ in the transfer function associated to $\mathcal{P}$ and to apply the Routh-Hurwitz stability test on the corresponding closed loop polynomials. The final test is then on a very large (but finite!) set of multilinear inequalities involving the real parameter λ and the parameters of the controller. Checking whether these inequalities can be simultaneously satisfied is then only a matter of a finite number of rational operations (addition, substraction, multiplication and division) on the coefficients appearing in the inequalities (this by Tarsky-Seidenberg, or elimination theory, see Anderson [5]).

Of course the procedure described above is totally inefficient (we definitely do not recommend it!) but it nevertheless shows that the stabilizability of $\mathcal{P}$ can be checked in a finite number of steps involving rational operations only, thus preparing the room for more effective techniques that also use only finitely many rational operations but that are simpler to implement.

This 'finite situation' appears because an upper bound on the degree of the controller can be given a-priori. The underlying infinite dimensional space for the controller can therefore be by-passed. The same argument can be used for almost any problem in system theory that involves control design with a controller of a-priori fixed degree.

The situation so depicted is different when considering simultaneous stabilization of a finite number of systems. Consider for example the subset of $\mathcal{P}$ defined by $\mathcal{P}' = \{k_i p(s) : i = 1, 2, 3 : k_1 = 0, k_2 = \frac{1}{2}, k_3 = 1\}$ where $p(s)$ is as in $\mathcal{P}$. It is known (see for example Ghosh [52]) that, except for the case $n = 1$, there exists no a-priori bound on the degree of a stabilizing controller of $\mathcal{P}$ that depends only on n. There are examples of systems $p(s)$ of degree 2 for which stabilizing controllers of $\mathcal{P}'$ exist but with a degree that needs to be, say, at least 1995. Therefore we are not ensured of the existence of a finite rational procedure to decide whether $\mathcal{P}$ is stabilizable. In this case the underlying infinite dimensional structure of the set of controllers can not be trivially by-passed and this is what makes the problem difficult.

The question of simultaneous stabilization of linear systems has been formulated for some years now. The first explicit statement was made in 1982 (see Saeks and Murray [102]). From this very beginning the question was clearly identified:

> Let p_1, p_2, ..., p_k be k scalar linear time invariant systems. Under what condition does there exists a controller c that is stabilizing for each p_i $(i = 1, ..., k)$?

We extract from the conclusion of the original paper of Saeks and Murray: "At the present time, no computationally feasible solution to the simultaneous stabilization problem is known, except in the case of two systems." More than a decade after this statement was made, and despite all research efforts, the same conclusion is still applicable: at the present time there exists no tractable condition for simultaneous stabilization of three or more linear systems. This question constitutes one of the most outstanding and challenging open problems in linear system theory and was the initiator of the research reported in this monograph.

Compared to the other three robust control techniques, simultaneous stabilization has received little attention from the control community. It is mentioned in only some of the standard books on control theory and it has given rise to the relatively small number of about one hundred or so research papers. We briefly review this literature here.

The case $k = 1$ –the stabilization of a single system– is easily dealt with. There always exists a stabilizing controller for a single system. Moreover, once a stabilizing controller of a single system is found, it is easy to parametrize the infinite set of all stabilizing controllers of this system. This parametrization is known as the Youla-Kucera parametrization and was discovered in 1976 by Youla *et al.* [124] and by Kucera [80]. The parametrization relies crucially on the powerful algebraic fractional description of linear systems that took several years to achieve its final form (see Vidyasagar [108], Pernebo [97], Youla [123] and many others).

By using the Youla-Kucera parametrization, it is possible to rephraze simultaneous stabilization of two systems into strong stabilization –stabilization with a stable controller– of a single system. This particularity was discovered in 1982 for scalar (single input single output) systems by Saeks and Murray [102] and by Vidyasagar and Viswanadham [110] for multivariable systems. The strong stabilization question had then been solved for almost then years. In

1974 Youla, Bongiorno and Lu [123] had published a surprising and elegant answer to the strong stabilization question: a system is stabilizable by a stable controller if and only if it has an even number of real unstable zeros between each pair of real unstable poles. What is really remarkable about this condition –known as the *parity interlacing property*– is that it involves only *real* poles and zeros. Thus the simultaneous stabilizability question for $k = 2$ is fully solved: translated into a strong stabilization question of a related system and by using the Youla-Kucera parametrization it can then be checked by testing the parity interlacing property –a tractable necessary and sufficient condition.

For $k = 3$ the picture is different. Presently available results are either in the form of necessary conditions, sufficient conditions or untractable necessary and sufficient conditions i.e. equivalent formulations.

Necessary conditions have been given by Ghosh [50], Wei [118,119] and Blondel, Gevers, Mortini and Rupp [16]. These conditions all rely on the same underlying idea as that of the parity interlacing property; they all involve conditions on the real axis only.

Sufficient conditions have been given by Maeda and Vidyasagar [90], Alos [3], Emre [39], Kwakernaak [81], Wei [116,117], Debowsky and Kurylowicz [34], Blondel, Campion and Gevers [14] and Blondel [13]. The results contained in these papers cover cases that are of important practical significance. A typical example is that of minimum phase systems (see Kwakernaak [81]): if k linear minimum phase systems have the same high frequency gain sign then they are simultaneously stabilizable. The minimum phase assumptions is non generic in a mathematical sense but it covers a situation of important practical significance.

Necessary and sufficient conditions are given, for example, in Vidyasagar and Viswanadham [110], Ghosh [50] and Blondel, Gevers, Mortini and Rupp [16]. A typical example of such necessary and sufficient conditions is contained in Vidyasagar [110]: from k systems p_i ($i = 1, ..., k$) it is possible to construct $k - 1$ systems p_i' ($i = 1, ..., k - 1$) in such a way that the k systems p_i are simultaneously stabilizable if and only if the $k - 1$ systems p_i' are simultaneously stabilizable with a *stable* controller. The equivalence is thus between simultaneous stabilization of k systems and simultaneous stabilization of $k - 1$ systems with a stable controller. In a certain sense such an equivalence relation solves the original question since it provides necessary and sufficient conditions for simultaneous stabilization. These conditions are however of no practical help since there exists no tractable criterion to decide whether two or more systems are simultaneously stabilizable with a stable controller.

There exist at present no tractable necessary and sufficient conditions for testing the simultaneous stabilizability of three or more linear systems.

It is shown in Vidyasagar, Levy and Viswanadham [114] that, for the graph topology, k multivariable $n \times m$ systems are generically stabilizable when $k \leq \min(m, n)$. Generically it is possible to simultaneously stabilize as many multivariable systems as there are inputs or outputs to each systems, whichever is larger. Except for genericity results of the same kind the features that are specific to simultaneous stabilization are almost all contained in the scalar case and we shall restrict our attention to such systems. However, several contributions are stated in the framework of multivariable systems, see e.g. Vidyasagar and Viswanadham [110].

Simultaneous stabilization can also be treated in a time domain context. An analysis of simultaneous stabilization in the state space context is given in Minto and Vidyasagar [91] and in Vidyasagar [112]. The example contained in [91] is of a GE-21 jet engine, a 7×7 non linear system, that is stabilized at 6 different operating points with the same linear controller.

For completeness of this tour d'horizon of the literature let us mention that we are not aware of any analysis of simultaneous stabilization in a non linear context but that linear systems are always simultaneous stabilizable by time varying controllers (see Kabamba [72] and references therein).

1.2 Organization

This monograph focuses on the question of simultaneous stabilization of scalar linear systems:

> Let p_1, p_2, ..., p_k be k scalar linear systems. Under what condition does there exists a controller c that is stabilizing for each p_i $(i = 1, ..., k)$?

We are mostly interested by the existence question and we make only few remarks on constructive procedures.

The volume consists of 6 chapters and 3 appendices. The chapters are as independent as possible of each others. They all start with a short introduction and end with a concluding section that summarizes the content and gives bibliographical references. The first two appendices contain short introductions to the concepts of rings, Banach algebra and analytic functions. The third appendix presents results on the range of analytic functions. A list of references and a list of notations are provided at the end of the volume.

We now briefly summarize the content of each chapter.

Chapter 2: Stabilization

This chapter contains all the basic definitions: stabilization, strong, bistable and simultaneous stabilization, etc. There exist several conceptual ways to look at stabilization and we present two of these: the algebraic factorization approach and the geometrical avoidance interpretation. Both approaches are used in the later chapters. The geometrical interpretation of stabilization underlines many of the results contained in the monograph.

Chapter 3: Youla-Kucera parametrization

The Youla-Kucera parametrization of the set of all stabilizing controllers of a given system is the stepping stone to many control design strategies. It is here presented and used to derive equivalences between strong, bistable and simultaneous stabilization. This chapter also contains general comments on properness and finiteness of controllers.

Chapter 4: Necessary conditions

A rational function that has no poles in the right half plane has no poles on the positive real axis. This elementary observation is the guideline used in the chapter to obtain necessary conditions for simultaneous stabilization. If all the closed loop transfer functions associated to the systems p_i ($i = 1, ..., k$) in feedback with a controller c have no poles in the right half plane, then this controller is such that all the closed loop transfer functions have no poles on the positive real axis. Such a controller is said to be $R_{+\infty}$-stabilizing for p_i ($i = 1, ..., k$). The existence of an $R_{+\infty}$-stabilizing controller is thus a necessary condition for the existence of a stabilizing controller. In this chapter we derive tractable necessary and sufficient conditions for simultaneous $R_{+\infty}$-stabilizability of k systems and for strong and bistable $R_{+\infty}$-stabilization. These conditions encompass all known necessary conditions and are expressed in the form of interlacement properties on poles and zeros of the systems and of related rational functions.

Chapter 5: Sufficient conditions

In this chapter we show that the necessary conditions derived in Chapter 4 are also sufficient for simultaneous stabilization of two systems and for strong stabilization. Then we show that this property does not flow on to bistable stabilization or to simultaneous stabilization of more than two systems: three systems that are simultaneously $R_{+\infty}$-stabilizable are not guaranteed simultaneous stabilizable. The conditions given in Chapter 4 for bistable or simultaneously stabilization of 3 or more systems are not sufficient. The proof of this result makes a crucial use of results from analytic function theory. In the last two sections we provide examples of sufficient conditions that are related to the avoidance concept given in Chapter 2 and to H_∞ control design.

Chapter 6: Necessary and sufficient conditions

The result contained in this closing chapter is that the simultaneous stabilization question is rationally undecidable: it is not possible to find necessary and sufficient conditions for simultaneous stabilization of three or more systems that involve only a finite combination of rational operations (additions, substractions, multiplications and divisions), logical operations ('and' and 'or') and sign test operations (equal to, greater than, greater than or equal to, etc) on the coefficients of the three systems. This chapter extinguishes a line of investigation by pointing to an inherent limitation of simultaneous stabilization conditions: necessary and sufficient conditions must be in a form that involve more than just the above described elementary operations.

Chapter 2

Stabilization

Trying to express complex issues in very simple language
is an excellent exercise in discovering
how well you really understand them.

S. Birchmore, New Scientist.

2.1   Introduction

In this chapter we give the precise definitions of stabilization and simultaneous, strong and bistable stabilization. Each of these notions can be expressed both in an algebraic and in a geometric framework. We introduce these two complementary approaches in Sections 3 and 4.

The concept of stability that is used in most parts of this monograph is the usual one of continuous time stability: a rational function is stable if and only if it has no poles in the extended closed right half plane. In Section 5 we present an extension of this notion of stability to other subsets of the complex plane. This generalised concept of stability will be used in later chapters.

A short summary and bibliographical references are given in Section 6.

2.2 Definitions

We consider finite scalar systems that are linear and time invariant. By *plants* or *systems* we understand systems that satisfy all these assumptions. Such systems are represented in the frequency domain by real rational functions. The set of real rational functions (or, for short, the set of rational functions) is denoted by $\mathbf{R}(s)$ and the extended complex plane $\mathbf{C} \cup \{\infty\}$ is denoted by $\mathbf{C}_\infty$.

Poles and zeros of rational functions are defined as usual. Assume that $p(s) \in \mathbf{R}(s)$, then $p(s)$ has a *pole* at $s_0 \in \mathbf{C}$ if and only if $\lim_{s \to s_0} p(s) = \infty$ and $p(s)$ has a zero at $s_0 \in \mathbf{C}$ if and only if $\lim_{s \to s_0} p(s) = 0$. The rational function $p(s)$ has a pole at infinity if and only if $\lim_{s \to \infty} p(s) = \infty$ and it has a zero there if and only if $\lim_{s \to \infty} p(s) = 0$. These last two conditions are satisfied when the degree of the numerator is, respectively, strictly larger than or strictly smaller than that of the denominator. Except for $p(s) \equiv 0$ that has zeros everywhere in $\mathbf{C}_\infty$ and whose degree is defined to be equal to zero, rational functions have an equal finite number of poles and zeros in $\mathbf{C}_\infty$ (when counting poles and zeros we always assume that multiplicities are taken into account). This number is the *degree* of $p(s)$ and it coincide with the usual definition of the degree of a polynomial.

A rational function is *proper* if it has no poles at infinity. It is *strictly proper* if it has a zero at infinity and it is *biproper* if it has neither poles nor zeros at infinity (the reason for this terminology is due to an analogy with the set of rational numbers: a rational number is proper if the magnitude of its numerator does not exceed that of its denominator, polynomials are ordered by degree, hence the name).

When a rational function is strictly proper, its *relative degree* is equal to the multiplicity of the zero at infinity. Alternatively, it is the difference between the denominator degree and the numerator degree.

A rational function is *stable* if it is proper and has no poles in $C_+ = \{s \in C : \Re(s) \geq 0\}$ or, in condensed form, if it has no poles in $C_{+\infty} = \{s \in C : \Re(s) \geq 0\} \cup \{\infty\}$. The poles and zeros of $p(s)$ in $C_{+\infty}$ are the unstable poles and zeros of p. The subset $C_{+\infty}$ can be thought of as the *instability region*. The *stability region* is then the complement of $C_{+\infty}$ in C_{∞}. A rational function that is not stable is *unstable*.

A rational function $p(s)$ will be called *inverstable* if it has no zeros in $C_{+\infty}$ or, alternatively, if $p(s)^{-1}$ is stable. If $p(s)$ is both stable and inverstable it is called *bistable*.

A rational function is *minimum phase* if it has no zeros in $\mathbf{C}_+ = \{s \in \mathbf{C} : \Re(s) \geq 0\}$. All inverstable rational functions are minimum phase but the converse is not true: $p(s) = \frac{s+1}{s^2-3s+1}$ is minimum phase but is not inverstable.

Throughout the monograph we consider a *controller* $c(s) \in \mathbf{R}(s)$ to be within a unity feedback loop with the system $p(s) \in \mathbf{R}(s)$ (see Figure 2.1).

It is possible to choose other system-controller configurations but they all lead to the same conceptual ideas. For notational convenience we chose a negative sign for the unity feedback so that a positive sign appears in the expression of the closed loop transfer functions. With this configuration the four transfer functions between x_1, x_2 and y_1, y_2 can be computed as

$$y_1 = \frac{c(s)}{1 + p(s)c(s)} x_1,$$

$$y_2 = \frac{p(s)c(s)}{1 + p(s)c(s)} x_1,$$

$$y_1 = \frac{1}{1 + p(s)c(s)} x_2,$$

and

$$y_2 = \frac{p(s)}{1 + p(s)c(s)} x_2.$$

The controller $c(s)$ *externally stabilizes* the system $p(s)$ if and only if $\frac{p(s)c(s)}{1+p(s)c(s)}$ is stable. The same controller *internally stabilizes* the system $p(s)$ if and only if the four transfer functions between x_1, x_2 and y_1, y_2 are stable.

An internally stabilizing controller is also externally stabilizing but the converse is not true: the controller $c(s) = \frac{s-1}{s+1}$ externally stabilizes the system $p(s) = \frac{s}{s-1}$ but it does not internally stabilize

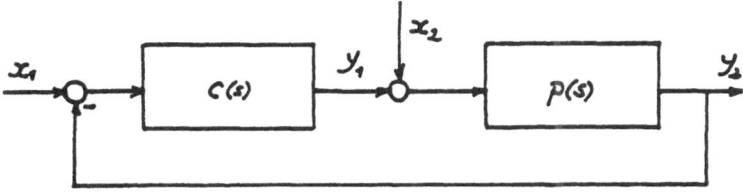

Figure 2.1: The closed loop feedback configuration with the controller $c(s)$ and the system $p(s)$.

$p(s)$ since $\frac{pc}{1+pc} = \frac{s}{2s+1}$ is stable whereas $\frac{p}{1+pc} = \frac{s(s+1)}{(2s+1)(s-1)}$ is not (when clear from the context we drop the reference to the complex variable s).

The constraint for a controller to be internally and not just externally stabilizing corresponds to the usage to avoid unstable pole-zero cancellations between the system and the controller.

Theorem 2.1 *The controller $c \in \mathbf{R}(s)$ internally stabilizes the system $p \in \mathbf{R}(s)$ if and only if c is an externally stabilizing controller of p and the unstable zeros (respectively poles) of p are not poles (respectively zeros) of c.*

Proof Assume for sufficiency that c is an externally stabilizing controller of p and that p and c have no unstable pole-zero cancellations. By definition $\frac{pc}{1+pc}$ is stable. Also the rational function q defined by $q = \frac{pc}{1+pc}$ has unstable zeros at the unstable zeros of

p since these do not cancel with the unstable poles of c. Consequently, $\frac{q}{p} = \frac{c}{1+pc}$ is stable, as required. A similar argument shows that $\frac{p}{1+pc}$ and $\frac{1}{1+pc}$ are also stable.

For necessity, assume by contradiction that $\frac{pc}{1+pc}$ is stable and that one of the unstable poles of p cancels with a corresponding zero of c. Then, dividing the above expression by c yields a transfer function $\frac{p}{1+pc}$ that is unstable. A contradiction is achieved and necessity is proved. ∎

The objective of internal stability (and not just external stability) is recognized in linear system theory as one of the key issues in the design of a controller. In this volume we use only internal stabilization and write, for short, stabilization.

The concept of stabilization is symmetric in terms of the rational functions p and c. The controller c stabilizes the system p if and only if the 'controller' p stabilizes the 'system' c. It is often useful to keep this property of symmetry in mind. For example, the parametrization of the set of all the controllers that stabilize a given system (see next chapter) can also be seen, by symmetry, as a parametrization of all the systems that are stabilized by a given controller.

All our definitions are valid for finite real rational functions, thus excluding the possibility of having a system or a controller that is identically equal to infinity. In several theoretical situations however the treatment of infinite systems and controllers appears naturally and we will then allow this possibility.

The *infinite rational function* is denoted by ∞. It has no zeros but has poles everywhere in $\mathbf{C}_\infty$. It is thus non proper, minimum phase, unstable and inverstable. The inverse of ∞ is defined to be equal to 0.

The infinite controller stabilizes the system $p \in \mathbf{R}(s)$ if and only if p is inverstable. By symmetry, the controllers that stabilize the infinite system are exactly those that are inverstable. In particular $c = \infty$ stabilizes $p = \infty$. In the sequel we denote the set of *extended rational functions* $\mathbf{R}(s) \cup \{\infty\}$ by $\mathbf{R}'(s)$.

A system is *strongly stabilizable* if it is stabilizable by a stable controller and it is *bistably stabilizable* if it is stabilizable by a bistable controller. A system is not always strongly nor bistably stabilizable and the *strong and bistable stabilization problems* are those of finding necessary and sufficient conditions for a system to be strongly or bistably stabilizable.

Assume that $p_i \in \mathbf{R}'(s)$ $(i = 1, \ldots, k)$ are extended rational functions. The k systems p_i are *simultaneously stabilizable* if and only if there exists a controller $c \in \mathbf{R}'(s)$ that stabilizes all p_i $(i = 1, ..., k)$. Such a controller is a *simultaneous stabilizing controller* of p_i. Systems are not always simultaneously stabilizable and the *simultaneous stabilization problem of k systems* is one of finding necessary and sufficient conditions for k systems to be simultaneously stabilizable.

2.3 Algebraic framework

2.3.1 Introduction

The set of rational functions $\mathbf{R}(s)$ has the algebraic structure of a field. This is not true of the set of stable rational functions because a stable rational function is not always stably invertible: the rational function $\frac{s-1}{s+1}$ is stable but has no inverse in the set of stable rational functions. The adequate structure for the description of the set of stable rational functions is that of a ring.

In the first part of this section we analyse how the general notions associated with rings apply to the ring of stable rational functions.

Thereafter we use these notions to express stabilization and strong, bistable and simultaneous stabilization under the form of equations in the ring of stable functions. The main idea for doing this is the so-called *factorization approach*. A rational function can be factorized as a ratio of two polynomials but can also be factorized as a ratio of two *stable* rational functions and it is this factorization that is used in most parts of the monograph.

Definitions and preliminary results on rings and matrix rings are provided in Appendix A.

2.3.2 Ring of stable rational functions

It is a standard usage to denote the set of rational functions with no poles in some subset Λ of the complex plane by $\mathbf{R}_\Lambda(s)$ (see for example Pernebo [97] where such rational functions are called Λ-polynomials). With this notation the set of stable rational functions would be denoted by $\mathbf{R_{C+\infty}}(s)$ but for obvious conciseness reasons we shall refer to it with a shorthand notation. We denote the set of stable rational functions by S (i.e. $S = \mathbf{R_{C+\infty}}(s)$) and the set of bistable rational functions by U (S and U stand for Stable and Units).

S is a commutative ring for which the notions of inversion, division and coprimeness take the following forms.

Theorem 2.2 *Assume that $q, q_1, q_2 \in S$. Then*

1. *q is invertible if and only if q has no unstable zeros,*

2. *q_1 divides q_2 if and only if every unstable zero of q_1 is also, multiplicity included, an unstable zero of q_2,*

3. *q_1 and q_2 are coprime if and only if they have no common unstable zeros.*

A consequence of the characterization of invertible elements is that the set of invertible elements of the ring S (also called units of S) is the set U of bistable rational functions.

The ring S has an additional property: it is a principal ideal domain whose field of fractions is $\mathbf{R}(s)$ (see appendix A and [97]). Using the properties of principal ideal domains, we have:

Theorem 2.3 *Assume that $q \in \mathbf{R}(s)$, then there exist $n_q, d_q \in S$ and $x, y \in S$ that are such that*

1. *n_q and d_q are coprime in S (i.e. they have no common unstable zeros),*

2. *$n_q x + d_q y = 1$,*

3. $q = \frac{n_q}{d_q}$.

A factorization $\frac{n_q}{d_q}$ of a rational function $q \in \mathbf{R}(s)$ that satisfies these three properties is called a *coprime fractional factorization* of q in S or, for short, a *fractional factorization* of q in S.

The infinite rational function $q = \infty \in \mathbf{R}'(s)$ admits the fractional factorization $d_q = 0, n_p = u \in U$ where u is any bistable rational function. For this factorization we have $n_q \frac{1}{u} + d_q r = 1$ for any $r \in S$.

Fractional factorizations of rational functions in S are non unique: if $q = \frac{s-1}{s-2}$ then $n_q = \frac{s-1}{s+1}$ and $d_q = \frac{s-2}{s+1}$ both belong to S, are coprime, satisfy $n_q \frac{3}{2} - d_q \frac{1}{2} = 1$ and are such that $q = \frac{n_q}{d_q}$. The same is true for $n_q = \frac{s-1}{s+2}$ and $d_q = \frac{s-2}{s+2}$ for which $n_q \frac{4}{3} - d_q \frac{1}{3} = 1$. A characterization of all possible fractional factorizations of a given rational function is given in the next theorem.

Theorem 2.4 *Let $q \in \mathbf{R}(s)$. Then*

1. *if $\frac{n_q}{d_q}$ is a fractional factorization of q in S then any other fractional factorization of q in S is of the form $\frac{n_q'}{d_q'}$ where $n_q' = n_q u$ and $d_q' = d_q u$ for some $u \in U$,*

2. *if q is stable then $n_q = q \in S$, $d_q = 1$ is a fractional factorization of q in S,*

3. *if q is bistable then $n_q = q \in U$, $d_q = 1$ and $n_q = 1$, $d_q = \frac{1}{q} \in U$ are both fractional factorizations of q in S.*

Proof We examine the first case only, the other two are immediate consequences. Note first that, if $\frac{n_q}{d_q}$ is a fractional factorization of q in S, then $\frac{n_q u}{d_q u}$ is also a fractional factorization of q in S. Indeed,

$q = \frac{n_q}{d_q} = \frac{n_q u}{d_q u}$, $n_q u$ and $d_q u$ are coprime and, if $n_q x + d_q y = 1$ then $n_q u x' + d_q u y' = 1$ for $x' = \frac{x}{u} \in S$ and $y' = \frac{y}{u} \in S$.

It remains to show that all fractional factorizations are of this form. Assume therefore that $\frac{n_q}{d_q} = \frac{n_q'}{d_q'}$ for some n_q', $d_q' \in S$. Then $n_q d_q' = d_q n_q'$. But since n_q and d_q are coprime this last equality implies that n_q' divides n_q. n_q similarly divides n_q' and thus $n_q' = n_q u$ for some $u \in U$ as requested. ∎

The operation of factorization of a rational function into a ratio of two stable rational functions is the cornerstone of the *factorization approach* in system theory. We extract from the preface of the book *Controller System Synthesis: A Factorization Approach, M. Vidyasagar [108]* :

> The central idea [...] is that of "factoring" the transfer
> matrix of a (not necessarily stable) system as the "ra-
> tio" of two stable rational matrices. [...] It turns out
> that this seemingly simple stratagem leads to conceptu-
> ally simple and computationally tractable solutions to
> many important and interesting problems.

2.3.3   Ring concepts and stabilization

A controller stabilizes a system if and only if the four transfer functions associated to the system in closed loop with the controller are stable, i.e. if they all belong to the ring S. In this subsection we derive a more compact form for this condition by using the factorization approach.

Theorem 2.5 *Assume that $p, c \in \mathbf{R}(s)$ and let $\frac{n_p}{n_p}, \frac{n_c}{d_c}$ be arbitrary fractional factorizations of p and c in S. Define $u = n_p n_c + d_p d_c \in$*

S. Then c stabilizes p if and only if $u \in U$.

Proof Note first that, if $p = \frac{n_p}{d_p}$ and $c = \frac{n_c}{d_c}$, then

$$(pc + 1)^{-1} = \frac{d_p d_c}{n_p n_c + d_p d_c} = \frac{d_p d_c}{u},$$

$$p(1 + pc)^{-1} = \frac{n_p d_c}{n_p n_c + d_p d_c} = \frac{n_p d_c}{u},$$

$$c(1 + pc)^{-1} = \frac{d_p n_c}{n_p n_c + d_p d_c} = \frac{d_p n_c}{u},$$

$$\text{and} \qquad pc(1 + pc)^{-1} = \frac{n_p n_c}{n_p n_c + d_p d_c} = \frac{n_p n_c}{u}.$$

Clearly, if $u \in U$ then $\frac{1}{u} \in U$ and all the above transfer functions are stable. Sufficiency is thus proved.

It remains to prove that if the four transfer functions are all stable then $u = n_p n_c + d_p d_c \in U$. We therefore use the fact that n_p, d_p and n_c, d_c are coprime so that, by Theorem 2.3, there exists $x_p, y_p \in S$ and $x_c, y_c \in S$ such that $n_p x_p + d_p y_p = 1$ and $n_c x_c + d_c y_c = 1$. Check then that

$$x_c(x_p n_p n_c u^{-1} + y_p d_p n_c u^{-1}) + y_c(x_p n_p n_c u^{-1} + y_p d_p d_c u^{-1}) = u^{-1}.$$

Since all the functions involved in this expression are stable this implies that

$$u^{-1} \in S.$$

Thus $u \in U$ and the theorem is proved. ∎

By using this result we rewrite strong and bistable stabilization under the form of equations over S.

Theorem 2.6 *Assume that $p \in \mathbf{R}(s)$ and let $\frac{n_p}{d_p}$ be any fractional factorization of p in S. Then the following are equivalent:*

1. p is strongly stabilizable,

2. there exists $c \in S$ such that

$$n_p c + d_p \in U,$$

3. there exist $c \in S$ and $u \in U$ such that

$$n_p c + d_p u \in U,$$

4. there exist $n_c, d_c \in S$ such that

$$\begin{pmatrix} 0 & 1 \\ n_p & d_p \end{pmatrix} \begin{pmatrix} n_c \\ d_c \end{pmatrix} = \begin{pmatrix} u_1 \\ u_2 \end{pmatrix} \text{ for some } u_1, u_2 \in U,$$

5. there exist $n_c, d_c \in S$ such that

$$\begin{pmatrix} 0 & 1 \\ n_p & d_p \end{pmatrix} \begin{pmatrix} n_c \\ d_c \end{pmatrix} = \begin{pmatrix} 1 \\ u \end{pmatrix} \text{ for some } u \in U.$$

A similar formulation for bistable stabilization is:

Theorem 2.7 *Assume that $p \in \mathbf{R}(s)$ and let $\frac{n_p}{d_p}$ be any fractional factorization of p in S. Then the following are equivalent:*

1. p is bistably stabilizable,

2. there exists $c \in U$ such that

$$n_p c + d_p \in U,$$

3. there exist $n_c, d_c \in S$ such that

$$\begin{pmatrix} 0 & 1 \\ 1 & 0 \\ n_p & d_p \end{pmatrix} \begin{pmatrix} n_c \\ d_c \end{pmatrix} = \begin{pmatrix} u_1 \\ u_2 \\ u_3 \end{pmatrix} \text{ for some } u_1, u_2, u_3 \in U,$$

4. there exist $n_c, d_c \in S$ such that

$$
\begin{pmatrix} 0 & 1 \\ 1 & 0 \\ n_p & d_p \end{pmatrix} \begin{pmatrix} n_c \\ d_c \end{pmatrix} = \begin{pmatrix} 1 \\ u_2 \\ v_3 \end{pmatrix} \quad \text{for some } u_2, u_3 \in U.
$$

Simultaneous stabilization can also be written in terms of a matrix equation over S.

Theorem 2.8 *Assume that $p_i \in R(s)$ ($i = 1, ..., k$) and let $\frac{n_i}{d_i}$ be any fractional factorizations of p_i in S. Then there exists a simultaneous stabilizing controller for p_i ($i = 1, \ldots, k$) if and only if there exist $n_c, d_c \in S$ that are such that*

$$
\begin{pmatrix} n_1 & d_1 \\ n_2 & d_2 \\ \vdots & \\ n_k & d_k \end{pmatrix} \begin{pmatrix} n_c \\ d_c \end{pmatrix} = \begin{pmatrix} u_1 \\ u_2 \\ u_3 \\ \vdots \\ u_k \end{pmatrix} \quad \text{for some } u_i \in U \; (i = 1, \ldots, k)
$$

or such that

$$
\begin{pmatrix} n_1 & d_1 \\ n_2 & d_2 \\ \vdots & \\ n_k & d_k \end{pmatrix} \begin{pmatrix} n_c \\ d_c \end{pmatrix} = \begin{pmatrix} 1 \\ u_2 \\ u_3 \\ \vdots \\ u_k \end{pmatrix} \quad \text{for some } u_i \in U \; (i = 2, \ldots, k).
$$

Proof The systems p_i ($i = 1, \ldots, k$) are simultaneously stabilizable if and only if there exist $n_c, d_c \in S$ for which $n_i n_c + d_i d_c = u_i \in U$ ($i = 1, \ldots, k$). This is the first matrix form.

Redefining $n_c = n_c u_1^{-1}$ and $d_c = d_c u_1^{-1}$ in this first matrix condition, we obtain the second condition. ∎

Comparing the last three theorems, we see that strong stabilization and bistable stabilization are special cases of the simultaneous stabilization questions for two and three systems. They correspond to the choice of special systems of fractional factorization $n_p = 0$, $d_p = 1$ and $n_p = 1$, $d_p = 0$, i.e. they correspond to the choice of the systems $p = 0$ and $p = \infty$. A system p is strongly stabilizable if and only if the two systems p and 0 are simultaneously stabilizable and p is bistably stabilizable if and only if the three systems p, 0 and ∞ are simultaneously stabilizable. The same conclusion will appear in the next section where we provide a geometrical interpretation of this result.

2.4 Geometrical framework

2.4.1 Introduction

The formalism that is most commonly used to express stabilization questions is the algebraic factorization approach described in the previous section. This approach allows a transformation of strong, bistable and simultaneous stabilization into elegant sets of algebraic equations involving unknown stable rational functions.

In this section we give a geometrical line of attack of these questions. The underlying idea is the following: an extended rational function $q \in \mathbf{R}'(s)$ can be seen as a function that goes from $\mathbf{C}_\infty$ to $\mathbf{C}_\infty$. To each point $s \in \mathbf{C}_\infty$ corresponds a value $q(s) \in \mathbf{C}_\infty$. This provides a geometrical illustration of the behaviour of extended rational functions that can be used to interpret stabilization in geometrical terms.

Two extended rational functions intersect at some point s_0 in the extended complex plane if they take the same value there. In accordance with this concept of intersection we define the notion of avoidance –non intersection– between extended rational functions and we express stabilization in terms of avoidance.

A major advantage of looking at stabilization in this way is that it provides a visual support for intuition. This will appear in a crucial fashion in Chapter 4. Most of the proofs contained in that chapter are constructed from geometrical considerations and they would be correct but not easily understandable if a geometrical representation was not provided in parallel with the formal proof.

2.4.2 Intersection and avoidance

The two extended rational functions q_1 and q_2 *intersect* at the point $s_0 \in \mathbf{C}_\infty$ if and only if $q_1(s_0) = q_2(s_0)$. In accordance with this definition we say that $q_1(s)$ *avoids* $q_2(s)$ on $\Lambda \subset \mathbf{C}_\infty$ if and only if $q_1(s)$ and $q_2(s)$ have no intersections in Λ, that is, if and only if there exists no s_0 in Λ for which $p_1(s_0) = p_2(s_0)$.

Definition 2.1 (Intersection) *Assume that $q_1(s)$, $q_2(s) \in \mathbf{R}'(s)$. The points of intersection (for short, the intersections) of $q_1(s)$ and $q_2(s)$ in $\mathbf{C}_\infty$ are the points $s_0 \in \mathbf{C}_\infty$ for which $\lim_{s \to s_0} q_1(s) = \lim_{s \to s_0} q_2(s)$. The k extended rational functions $q_i(s) \in \mathbf{R}'(s)$ $(i = 1, ..., k)$ simultaneously intersect at the point $s_0 \in \mathbf{C}_\infty$ if they all take the same value at s_0, i.e. if $\lim_{s \to s_0} q_i(s) = \alpha$ for some $\alpha \in \mathbf{C}_\infty$ and all i with $1 \le i \le n$.*

Note that we consider extended rational functions in the definition. The points of intersection between a rational function $q \in \mathbf{R}(s)$ and the infinite rational function are the poles of q.

Because rational functions have polynomials of finite degree both in the numerator and in the denominator, the number of points of intersection between two distinct rational functions is finite. If n_1 and n_2 are the degrees of the two rational functions q_1 and q_2 then there are exactly $n_1 + n_2$ intersections between q_1 and q_2 in $\mathbf{C}_\infty$.

One of the main reasons for the introduction of the notion of intersection is the interest of the concept of non-intersection, i.e. the concept of avoidance. Two rational functions avoid each other in some subset Λ of the complex plane if they have no intersections in Λ.

Definition 2.2 (Avoidance) *Assume that $q_1, q_2 \in \mathbf{R}'(s)$ and that Λ is a subset of the extended complex plane $\mathbf{C}_\infty$. Then q_1 avoids q_2 in Λ if and only if q_1 and q_2 have no points of intersection in Λ. That is, if and only if $q_1(s) \neq q_2(s)$ for all $s \in \Lambda$.*

The definition should not be misunderstood. The fact that $q_1(s_1) = q_2(s_2)$ for some $s_1, s_2 \in \Lambda$ does *not* imply that $p_1(s)$ and $p_2(s)$ intersect in Λ. They do intersect in Λ only when they take the same value *at the same point* in Λ.

We are mostly interested by avoidance between extended rational functions over specific subsets Λ of the extended complex plane. These have, in general, to be thought of as instability zones. The notion of stability used up to now is one of continuous time stability for which the instability zone is the extended right half plane $\mathbf{C}_{+\infty}$. Hence we are primarily interested by avoidance between extended rational functions over $C_{+\infty}$. It is however true that all the results presented in this chapter for $C_{+\infty}$ have counterparts for any subset of the extended complex plane. This is properly formalized in the next section.

The next theorem shows how to compute the points of intersection in $C_{+\infty}$ between two extended rational functions.

Theorem 2.9 *Assume that $q_1(s), q_2(s) \in R(s)$ and let $\frac{n_1(s)}{d_1(s)}$ and $\frac{n_2(s)}{d_2(s)}$ be fractional factorizations of $q_1(s)$ and $q_2(s)$ in S. Then the points of intersection between $q_1(s)$ and $q_2(s)$ in $C_{+\infty}$ are the zeros of $n_1(s)d_2(s) - n_2(s)d_1(s) \in S$ in $C_{+\infty}$. $q_1(s)$ and $q_2(s)$ avoid each other in $C_{+\infty}$ if and only if $n_1(s)d_2(s) - n_2(s)d_1(s) \in U$.*

Proof First assume that $q_1(s)$ intersects $q_2(s)$ at s_0, i.e. $\lim_{s \to s_0} \frac{n_1(s)}{d_1(s)}$ $= \lim_{s \to s_0} \frac{n_2(s)}{d_2(s)}$ for some $s_0 \in C_{+\infty}$. We show that, then, $(n_1 d_2 - n_2 d_1)(s_0) = 0$. Indeed

- if $d_1(s_0) = 0$ then $n_1(s_0) \neq 0$ since n_1 and d_1 are coprime but then $d_2(s_0) = 0$, and s_0 is a zero of $n_1 d_2 - n_2 d_1$,

- if $d_1(s_0) \neq 0$ then $d_2(s_0) \neq 0$ and then $\lim_{s \to s_0} n_1(s)d_2(s) = \lim_{s \to s_0} n_2(s)d_1(s)$ which implies that $(n_1 d_2 - n_2 d_1)(s_0) = 0$.

Suppose now that $(n_1 d_2 - n_2 d_1)(s_0) = 0$ for some $s_0 \in C_{+\infty}$. We show that $\lim_{s \to s_0} \frac{n_1(s)}{d_1(s)} = \lim_{s \to s_0} \frac{n_2(s)}{d_2(s)}$. Indeed

- if $d_1(s_0) = 0$ then $d_2(s_0) = 0$ since, n_1 and d_1 being coprime, we have $n_1(s_0) \neq 0$. This implies that q_1 and q_2 have a common pole at $s_0 \in C_{+\infty}$ and hence they intersect there,

- if $d_1(s_0) \neq 0$ then $d_2(s_0) \neq 0$ since otherwise $n_2(s_0)$ and $d_2(s_0)$ would both be equal to zero, which is impossible. Thus

$\left(\frac{n_1 d_2 - n_2 d_1}{d_1 d_2}\right)(s_0) = 0$, which implies that q_1 and q_2 intersect at s_0.

The theorem is proved. ∎

The next result is trivial and follows from the definitions. It nevertheless unveils the fundamental reason for the link between avoidance and stabilization.

Theorem 2.10 *Assume that $q \in \mathbf{R}'(s)$. Then q is stable if and only if it avoids ∞ in $\mathbf{C}_{+\infty}$.*

Proof The points of intersection between q and ∞ are the poles of q. q avoids ∞ in $\mathbf{C}_{+\infty}$ if and only if it has no poles there, i.e. if it is stable. ∎

We develop the link between avoidance and stability and between avoidance and stabilization in more detail in the next subsection. Before doing this we need a couple of technical lemmas. Remember that the inverse of the infinite rational function is equal to 0.

Lemma 2.1 *Assume that $q_1, q_2 \in \mathbf{R}'(s)$. Then q_1 avoids q_2 in $\mathbf{C}_{+\infty}$ if and only if $\frac{1}{q_1}$ avoids $\frac{1}{q_2}$ in $\mathbf{C}_{+\infty}$.*

Proof Assume by contradiction that q_1 avoids q_2 in $\mathbf{C}_{+\infty}$ and that $\frac{1}{q_1}$ intersects $\frac{1}{q_2}$ at some $s_0 \in \mathbf{C}_{+\infty}$. Define α by $\alpha = \frac{1}{q_1(s_0)} = \frac{1}{q_2(s_0)}$. α is either finite, in which case $q_1(s_0) = q_2(s_0)$, or α is infinite and then q_1 and q_2 have a common zero at s_0. In both cases q_1 and q_2 have a point of intersection at s_0. ∎

The avoidance property is thus preserved under inversion. It is tempting to think that the avoidance property is similarly preserved under addition: assume that $q_1 \in \mathbf{R}(s)$ avoids $q_2 \in \mathbf{R}(s)$ in

$C_{+\infty}$ and let $q_3 \in \mathbf{R}(s)$, does it then follow that $q_1 + q_3$ avoids $q_2 + q_3$ in $C_{+\infty}$? This is not true as the next example shows: $q_1 = 1$ certainly avoids $q_2 = 0$ in $C_{+\infty}$, but if $q_3 = \frac{1}{s-1}$ then $q_1 + q_3 = \frac{s}{s-1}$ does not avoids $q_2 + q_3 = \frac{1}{s-1}$ in $C_{+\infty}$ since they both have a pole at $s_0 = 1 \in C_{+\infty}$.

In the same vein avoidance is not preserved under multiplication. Consider for example $q_1 = 1, q_2 = 2$ and $q_3 = \frac{1}{s-1}$. Then q_1 avoids q_2 on $C_{+\infty}$ but $q_1 q_3$ does not avoid $q_2 q_3$ on $C_{+\infty}$ since they intersect at $s_0 = 1$.

Despite these disappointing conclusions –avoidance is not preserved under addition or multiplication– it is possible to adopt a more positive attitude. In the next result we provide conditions under which avoidance is preserved.

Lemma 2.2 *Assume that $q_1, q_2 \in \mathbf{R}(s)$. Then q_1 avoids q_2 in $C_{+\infty}$ if and only if the next two conditions are satisfied*

1. *q_1 and q_2 have no common poles in $C_{+\infty}$ and*

2. *$q_1 - q_2$ avoids 0 in $C_{+\infty}$.*

Proof For necessity assume that q_1 avoids q_2 in $C_{+\infty}$. Then they have no common poles in $C_{+\infty}$, since otherwise they would intersect at these common poles. Condition 1 is thus satisfied. But then q_1 and q_2 are never equal to infinity simultaneously in $C_{+\infty}$ and therefore the usual rules of limits apply. Namely, $\lim_{s \to s_0}(q_1(s) - q_2(s)) = \lim_{s \to s_0} q_1(s) - \lim_{s \to s_0} q_2(s)$ for any $s_0 \in C_{+\infty}$. But because q_1 avoids q_2 in $C_{+\infty}$ this means that $\lim_{s \to s_0} q_1(s) - \lim_{s \to s_0} q_2(s) \neq 0$ for all $s_0 \in C_{+\infty}$. Thus $\lim_{s \to s_0}(q_1(s) - q_2(s)) \neq 0$ for all $s_0 \in C_{+\infty}$ and $q_1 - q_2$ avoids 0 in $C_{+\infty}$ as requested by condition 2.

For sufficiency suppose that q_1 and q_2 have no common poles in $\mathbf{C}_{+\infty}$. Then, by the same argument as above, $\lim_{s \to s_0}(q_1(s) - q_2(s)) = \lim_{s \to s_0} q_1(s) - \lim_{s \to s_0} q_2(s)$. But by the second assumption, $q_1 - q_2$ avoids 0 in $\mathbf{C}_{+\infty}$. Thus $\lim_{s \to s_0}(q_1(s) - q_2(s))$ is never equal to 0 when $s_0 \in \mathbf{C}_{+\infty}$ and the conclusion follows. ∎

We can give a stronger version of this last lemma (substituting $q_3 = -q_1$ in the next lemma we get Lemma 2.2).

Lemma 2.3 *Assume that* $q_1, q_2, q_3 \in \mathbf{R}(s)$. *The following two statements are equivalent:*

1. *q_1 and q_2 have no common poles in $\mathbf{C}_{+\infty}$ and $q_1 + q_3$ avoids $q_2 + q_3$ in $\mathbf{C}_{+\infty}$,*

2. *$q_1 + q_3$ and $q_2 + q_3$ have no common poles in $\mathbf{C}_{+\infty}$ and q_1 avoids q_2 in $\mathbf{C}_{+\infty}$.*

Proof Throughout the proof when writing 'avoid' we understand 'avoid in $\mathbf{C}_{+\infty}$'.

$1 \Rightarrow 2$ Assume that $q_1 + q_3$ avoids $q_2 + q_3$. Then by the previous lemma, $q_1 + q_3$ and $q_2 + q_3$ have no common poles in $\mathbf{C}_{+\infty}$ and $(q_1 + q_3) - (q_2 + q_3) = q_1 - q_2$ avoids 0. But then, using the assumption that q_1 and q_2 have no common poles in $\mathbf{C}_{+\infty}$ and by a second application of Lemma 2.2, we get that q_1 avoids q_2. The two conditions have thus been proved.

$2 \Rightarrow 1$ Assume that q_1 avoids q_2. Then, by Lemma 2.2, $q_1 - q_2 = (q_1 + q_3) - (q_2 - q_3)$ avoids 0. By assumption, $q_1 + q_3$ and $q_2 + q_3$ have no common poles in $\mathbf{C}_{+\infty}$ and thus a second application of Lemma 2.2 leads to the desired fact that $q_1 + q_3$ avoids $q_2 + q_3$ in $\mathbf{C}_{+\infty}$. ∎

If $q_3(s)$ is stable and if $q_1(s)$ avoids $q_2(s)$ in $\mathbf{C}_{+\infty}$ then a direct application of Lemma 2.3 shows that $q_1(s) + q_3(s)$ avoids $q_2(s) + q_3(s)$ in $\mathbf{C}_{+\infty}$. Thus the avoidance property is preserved under addition of a *stable* rational function.

We can also obtain a result similar to Lemma 2.3 but for the case of multiplication of rational functions. Again we proceed in two steps and give a weak version first. The proofs are similar to those above.

Lemma 2.4 *Assume that $q_1, q_2 \in \mathbf{R}(s)$. Then q_1 avoids q_2 in $\mathbf{C}_{+\infty}$ if and only if the following two conditions are satisfied:*

1. *$\frac{q_1}{q_2}$ avoids 1 in $\mathbf{C}_{+\infty}$,*

2. *q_1 and q_2 have no common poles or zeros in $\mathbf{C}_{+\infty}$.*

We also have the stronger version.

Lemma 2.5 *Assume that $q_1, q_2, q_3 \in \mathbf{R}(s)$. Then $q_1 q_3$ avoids $q_2 q_3$ in $\mathbf{C}_{+\infty}$ if and only if the following two conditions are satisfied:*

1. *q_1 avoids q_2 in $\mathbf{C}_{+\infty}$,*

2. *$q_1 q_3$ and $q_2 q_3$ have no common poles in $\mathbf{C}_{+\infty}$.*

2.4.3   Avoidance and stabilization

The proof of the next theorem follows from the definition of avoidance.

Theorem 2.11 *Assume that $q \in \mathbf{R}(s)$. Then*

1. *q is stable if and only if q avoids ∞ in $\mathbf{C}_{+\infty}$,*

2. *q is inverstable if and only if q avoids* 0 *in* $C_{+\infty}$,

3. *q is bistable if and only if q avoids both* ∞ *and* 0 *in* $C_{+\infty}$.

Proof The first point is proved in Theorem 2.10, the second point can be similarly obtained and the third point is the combination of the first two. ∎

The concept of avoidance can also be used to determine when a controller stabilizes a system.

Theorem 2.12 *Assume that* $p, c \in R'(s)$. *Then c stabilizes p if and only if* $-\frac{1}{c}$ *avoids p in* $C_{+\infty}$.

Proof The case of an infinite system is easily dealt with. If $p = \infty$ then c stabilizes p if and only if c is inverstable, that is, if and only if $-\frac{1}{c}$ avoids $p = \infty$ in $C_{+\infty}$. Thus the theorem is proved for the infinite system. In the sequel we assume that p is finite.

For necessity assume that c stabilizes p. Then, by Theorem 2.1 we have the following two properties:

1. $\frac{pc}{1+pc}$ is stable and

2. there are no pole-zero cancellations between the unstable poles and zeros of p and c.

We use these two properties successively to prove the theorem.

1. Since $q = \frac{pc}{1+pc}$ is stable this means that q avoids ∞ in $C_{+\infty}$. By Lemma 2.1 this means that $\frac{1+pc}{pc} = \frac{1}{pc} + 1$ avoids 0 in $C_{+\infty}$. Since $\frac{1}{pc}$ and 1 have no common unstable poles we have, by Lemma 2.2, that $-\frac{1}{pc}$ avoids 1 in $C_{+\infty}$. We now use the second property to finish the proof of this part.

2. There are no cancellations between the unstable poles and zeros of p and c. Consequently, p and $-\frac{1}{c}$ have no common unstable poles or zero. But then, using Lemma 2.2, this implies that p avoids $-\frac{1}{c}$ in $\mathbf{C}_{+\infty}$ as requested.

For sufficiency assume that $-\frac{1}{c}$ avoids p in $\mathbf{C}_{+\infty}$. Then, by Lemma 2.2, p and $\frac{1}{c}$ have no common poles nor common zeros in $\mathbf{C}_{+\infty}$; alternatively, there are no unstable pole-zero cancellations between p and c.

But also by Lemma 2.2, $\frac{-1}{pc}$ avoids 1 in $\mathbf{C}_{+\infty}$. That is, $1 + \frac{1}{pc} = \frac{pc+1}{pc}$ avoids 0 in $\mathbf{C}_{+\infty}$ and thus $\frac{pc}{1+pc}$ avoids ∞ in $\mathbf{C}_{+\infty}$. In other words, $\frac{pc}{pc+1}$ is stable. We finish the proof by applying Theorem 2.1 that shows that, under these two conditions, c internally stabilizes p. ■

We are now able to restate stabilization and strong, bistable and simultaneous stabilization in terms of avoidance.

Theorem 2.13 *Assume that $p \in \mathbf{R}(s)$. Then p is*

1. *stabilizable*

2. *strongly stabilizable*

3. *bistably stabilizable*

if and only if there exist $q \in \mathbf{R}'(s)$ such that

1. *q avoids p in $\mathbf{C}_{+\infty}$*

2. *q avoids both p and 0 in $\mathbf{C}_{+\infty}$*

3. *q avoids $p, 0$ and ∞ in $\mathbf{C}_{+\infty}$*

In each case the controller defined by $c \triangleq -\frac{1}{q}$ has the desired properties.

Theorem 2.14 *Assume that $p_i \in \mathbf{R}'(s)$ $(i = 1, ..., k)$. Then p_i are simultaneously stabilizable if and only if there exists a rational function $q \in \mathbf{R}'(s)$ that avoids p_i in $\mathbf{C}_{+\infty}$ $(i = 1, ..., k)$. In such case the controller defined by $c = -\frac{1}{q}$ is the desired stabilizing controller.*

Again, the restatement so achieved shows that strong and bistable stabilization are special cases of simultaneous stabilization of two and three systems respectively.

At this stage the reader may be worried about two legitimate control theory preoccupations: the finiteness and the properness of the controller c. It is indeed desirable for practical purposes to obtain a controller that is not only finite but that is also proper, i.e. with no poles at infinity. In the theorem above, for example, these two requirements are not automatically satisfied: if $q = 0$ then $c = \infty$ and if q is proper then c is not. We shall prove in the next chapter (see Theorem 3.4 and Theorem 3.5) that if k systems (not necessarily proper nor strictly proper) are simultaneously stabilizable by a (not necessarily proper nor finite) controller, then they are also simultaneously stabilizable by a finite and proper controller. In other words, and as long as we are concerned only with the existence of stabilizing controllers, we do not have to worry about finiteness and properness issues. What can be done with infinite non proper controllers can also be done with finite proper ones.

2.5   General setting

We have started this chapter by defining the concept of stable rational function and we have chosen a continuous time definition: a rational function is stable if and only if it has no poles in the extended right half plane $C_{+\infty}$. In discrete time a rational function is stable if and only if it has no poles of modulus larger than or equal to one. That is, if and only if it has no poles in the extended annulus $\{\lambda \in C : |\lambda| \geq 1\} \cup \{\infty\}$.

These two subsets of the complex plane can be put in bijective correspondence by the well known bilinear transformation

$$s \rightarrow \lambda = \frac{s+1}{s-1}$$

that maps the extended right half plane into the extended annulus and whose inverse is

$$\lambda \rightarrow s = \frac{\lambda+1}{\lambda-1}.$$

In Chapters 4, 5 and 6 we will extensively use the closed unit disk $\overline{D}$ as a canonical instability region. The extended right half plane can be put in bijective correspondence with the closed unit disc $\overline{D}$ by

$$s \rightarrow z = \frac{s-1}{s+1}$$

whose inverse is

$$z \rightarrow s = \frac{1+z}{1-z}.$$

If we define $\sigma : C_\infty \rightarrow C_\infty$ by $\sigma(z) = \frac{1+z}{1-z}$ then the next result holds.

Theorem 2.15 *Assume that $p(s) \in R(s)$. Then $p(s)$ has no poles in the extended right half plane if and only if $p(\sigma(z)) \in R(z)$ has no poles in $\overline{D}$.*

This means that a rational function $p(s)$ is stable in a continuous time sense if and only if $p(\sigma(z))$ is stable in a $\overline{D}$ sense. Because simultaneous stabilization is expressed in terms of the stability of rational functions, we have the following equivalence.

Theorem 2.16 *Assume that $p_i \in \mathbf{R}(s)$ $(i = 1, ..., k)$. Then p_i are simultaneously stabilizable if and only if the systems defined by $p'_i(z) = p_i(\sigma(z)) \in \mathbf{R}(z)$ $(i = 1, ..., k)$ are simultaneously stabilizable in a $\overline{D}$-stability context. (See hereafter for a precise definition of $\overline{D}$-stabilization.)*

As a corollary we also have a correspondence between strong and bistable stabilization in continuous and $\overline{D}$-stability set-ups.

Theorem 2.17 *Assume that $p(s) \in \mathbf{R}(s)$. Then $p(s)$ is stabilizable by a controller with no poles (respectively with no poles nor zeros) in the extended right half plane if and only if the system defined by $p'(z) = p(\sigma(z)) \in \mathbf{R}(z)$ is $\overline{D}$-stabilizable by a controller with no poles (respectively with no poles nor zeros) in $\overline{D}$.*

Other regions of the complex plane can also be of practical interest in stabilization problems. We may consider for example a region that, in system theory jargon, guarantees exponential stability with a prescribed speed of decay. These different examples provide a motivation for the definition of a general notion of Λ-stability.

Let Λ be an arbitrary subset of the extended complex plane. A rational function is Λ-*stable* if it has no poles in Λ, it is Λ-*inverstable* if it has no zeros in Λ and it is Λ-*bistable* if it is both Λ-stable and Λ-inverstable. In accordance with our previous notations we denote by $S(\Lambda)$ and by $U(\Lambda)$ the sets of Λ-stable and Λ-bistable rational functions.

It can be checked that $S(\Lambda)$ is a ring that specializes to S when $\Lambda = \mathbf{C}_{+\infty}$, to $\mathbf{R}(s)$ when $\Lambda = \emptyset$, to $\mathbf{R}[s]$ when $\Lambda = \mathbf{C}$, to $\mathbf{R}$ when $\Lambda = \mathbf{C}_{\infty}$ and to the set of proper rational functions when $\Lambda = \{\infty\}$. All these different rings can be thought of as representatives of the same ring $S(\Lambda)$ for different choices of sets Λ.

The notions defined in the first section of this chapter carry over for Λ-stability. A controller $c \in \mathbf{R}'(s)$ Λ-*stabilizes* a system $p \in \mathbf{R}'(s)$ if and only if the four transfer functions associated to p in closed loop with c are Λ-stable. A system is *strongly Λ-stabilizable* if and only if it is Λ-stabilizable by a Λ-stable controller and it is *bistably Λ-stabilizable* if and only if it is Λ-stabilizable by a Λ-bistable controller. Finally, the k systems $p_i \in \mathbf{R}(s)$ $(i = 1, ..., k)$ are *simultaneously Λ-stabilizable* if and only if there exists a controller c that Λ-stabilizes all the systems p_i $(i = 1, ..., k)$.

If the subset Λ does not contain the whole extended real line $\mathbf{R}_{\infty}$ then the set $S(\Lambda)$ is a principal ideal domain whose field of fractions is $\mathbf{R}(s)$. The assumption that there exists some $x_0 \in \mathbf{R}_{\infty}$ with $x_0 \notin \Lambda$ is needed for $\mathbf{R}(s)$ to be the field of fractions of $S(\Lambda)$. Let for example $p(s) = \frac{1}{s-1} \in \mathbf{R}(s)$ and let $\frac{n_p(s)}{d_p(s)}$ be any fractional decomposition of $p(s)$ in $S(\Lambda)$. Then, either $n_p(s)$ or $d_p(s)$ have a pole on $\mathbf{R}_{\infty}$. Hence, if $\mathbf{R}_{\infty} \subset \Lambda$ it is impossible to factorize $p(s)$ in such a way that both $n_p(s)$ and $d_p(s)$ are members of $S(\Lambda)$.
We have the following main property of $S(\Lambda)$.

Theorem 2.18 *Assume that $\Lambda \subset \mathbf{C}_{\infty}$ and suppose that there exists $x_0 \in \mathbf{R}_{\infty}$ with $x_0 \notin \Lambda$. Then $S(\Lambda)$ is a principal ideal domain whose* *field of fractions is* $\mathbf{R}(s)$.

Proof It is shown in Pernebo [97] (see also Vidyasagar [108]) that $S(\Lambda)$ is an Euclidean ring for the degree function defined by

$$\delta_\Lambda : S(\Lambda) \setminus \{0\} \to \mathbf{N} : p(s) \to \text{ number of zeros of } p(s) \text{ in } \Lambda.$$

$S(\Lambda)$ is also a domain and, since Euclidean rings are principal ideal rings, the first point is proved.

Next, we prove that the field of fractions of $S(\Lambda)$ is $\mathbf{R}(s)$. Any rational function $p(s) = \frac{n(s)}{d(s)} \in \mathbf{R}(s)$ with $n(s), d(s) \in \mathbf{R}[s]$ and with no common zeros can be factorized as

$$p(s) = \left(\frac{n(s)}{(s - x_0)^m} \right) \left(\frac{d(s)}{(s - x_0)^m} \right)^{-1} = n_p(s)(d_p(s))^{-1}.$$

If m is equal to the largest polynomial degree of $n(s)$ and $d(s)$ then $n_p(s)$ and $d_p(s)$ are both members of $\mathbf{R}(s)$, have a unique pole at $x_0 \notin \Lambda$ and have no common zeros in $\mathbf{C}_\infty$. Therefore they form a coprime fractional factorization of $p(s)$ in $S(\Lambda)$ and thus the field of fractions of $S(\Lambda)$ is $\mathbf{R}(s)$. ∎

Some of the results contained in this monograph are stated for the ring S but need only the 'principal ideal domain' assumption. These results remain valid for arbitrary principal ideal domains and, in particular, they have a version for $S(\Lambda)$ with $\Lambda = \mathbf{R}_{+\infty}$. This will be used in Chapter 4.

2.6 Summary and bibliography

We have discussed two different interpretations of stabilization.

The algebraic approach analyses rational functions as ratios of stable rational functions. The geometric approach interprets rational functions as mappings from and to the extended complex plane.

Stabilization and strong, bistable and simultaneous stabilization can all be expressed with these two formalisms. In the first case as equations over the ring S and in the second case as avoidance of rational functions on subsets of the extended complex plane. In both cases the notions involved can be extended so as to encompass the general notion of Λ-stability.

The strong stabilization question was first addressed by Youla, Bongiorno and Lu in 1974 [123] and the simultaneous stabilization question was explicitly stated for the first time by Saeks and Murray [102], and by Vidyasagar and Viswanadham [110] for the multivariable case.

The proof that S is an Euclidean domain is given by Hung and Anderson [67]. See also Khargonekar and Ozgüler [75], Callier and Desoer [24] or Khargonekar and Sontag [76] for more properties of the ring S. The application of properties of the ring S to stabilization, i.e. the factorization philosophy, was initiated by Youla, Bongiorno and Lu [123]. Other references on this approach are Saeks and Murray [102], Vidyasagar [108,110] and Kucera [80].

The definition of avoidance and its application to stabilization can be found in Blondel, Campion and Gevers [18,17] and in Blondel, Gevers, Mortini and Rupp [16].

The use of the bilinear transformation to go from a discrete to a continuous time stability concept is common practice. The generalization of stability to Λ-stability is introduced in Pernebo [97]. Results for Λ-stability are contained for example in Vidyasagar [109], Wei [117] or Blondel [20].

Chapter 3

Youla-Kucera parametrization

3.1 Introduction

The simplest case of simultaneous stabilization, if this terminology is still appropriate in this simplified situation, is the one for which it is required to 'simultaneously' stabilize a single system.

A single system can always be stabilized by infinitely many controllers. In Section 2 we give a parametrization, the Youla-Kucera parametrization, of the set of all stabilizing controllers of a given system. The whole chapter centers around this parametrization.

In Sections 3 we provide a refinement of the parametrization for the cases where we want to exclude infinite or non proper controllers and we make some general comments on these particular issues.

In Section 4 we use the parametrization to derive equivalences be-

tween simultaneous stabilization questions. The simultaneous stabilization question for more than two systems is hard to answer
and this is probably the reason why many results on simultaneous
stabilization are given under the form of equivalences. Equivalent
formulations in simultaneous stabilization are transformations of
hard questions into equally difficult questions: the same problems
with different words. They nevertheless unveil links between seem-
ingly unrelated topics.

We first show the equivalence that exists between simultaneous stabilization of k systems and simultaneous stabilization of $k - 1$ systems by a stable controller and then give another similar equivalence
between simultaneous stabilization of k systems and simultaneous
stabilization of $k - 2$ systems by a bistable controller.

In a somewhat loose way, we may say that simultaneous stabilization of k systems, simultaneous stabilization of $k - 1$ systems with
a stable controller and simultaneous stabilization of $k - 2$ systems
with a bistable controller are all problems of the same level of dif-
ficulty.

3.2   The parametrization

The set of all stabilizing controllers of a given system $p \in \mathbf{R}'(s)$ is
defined by

$$Stab(p) \triangleq \{c \in \mathbf{R}'(s) : c \text{ stabilizes } p\}.$$

From Theorem 2.5 we know that if $p, c \in \mathbf{R}'(s)$ and $\frac{n_p}{d_p}, \frac{n_c}{d_c}$ are fractional factorizations of p and c in S, then c stabilizes p if and only
if

$$n_p n_c + d_p d_c \in U$$

and thus

$$Stab(p) = \{\frac{n_c}{d_c} : n_c, d_c \in S : n_p n_c + d_p d_c \in U\}.$$

For any given $p \in \mathbf{R}'(s)$ the set $Stab(p)$ is non empty. Indeed, the ring S is a principal ideal domain and hence if n_p, d_p are coprime in S then there exist $x, y \in S$ such that

$$n_p x + d_p y = 1 \in U.$$

Thus the system p is stabilized by the controller c defined by $c = \frac{x}{y}$.

Fractional factorizations of rational functions are non unique and this non uniqueness can be used to obtain another formulation of $Stab(p)$.

Assume that $p \in \mathbf{R}(s)$ and let $\frac{n_p}{d_p}$ be a factorization of p in S. If c is a stabilizing controller of p then any fractional factorization $\frac{n_c}{d_c}$ of c in S is such that $n_p n_c + d_p d_c = u \in U$. Consider one of these factorizations, say $c = \frac{n_c'}{d_c'}$, for which $n_p n_c' + d_p d_c' = u' \in U$. By Theorem 2.5 the controller c can then also be factorized by $c = \frac{n_c}{d_c}$ with $n_c = \frac{n_c'}{u'}$ and $d_c = \frac{d_c'}{u'}$. For this factorization we obtain $n_p n_c + d_p d_c = 1$ and thus our conclusion is:

Theorem 3.1 *Assume that $p \in \mathbf{R}(s)$ and let $\frac{n_p}{d_p}$ be a fractional factorization of p. Then*

$$
\begin{aligned}
Stab(p) &= \{\frac{n_c}{d_c} : n_c, d_c \in S : n_p n_c + d_p d_c \in U\} \\
&= \{\frac{n_c}{d_c} : n_c, d_c \in S : n_p n_c + d_p d_c = 1\}
\end{aligned}
$$

Proof See the discussion above. ∎

Thus, in order to find an expression for $Stab(p)$ it suffices to find all the solutions $n_c, d_c \in S$ of the equation $n_p n_c + d_p d_c = 1$ where n_p, d_p are given elements in S.

This question can be answered in a general ring context.

Theorem 3.2 *Assume that R is a ring and consider any two co-prime elements $a, b \in R$.*

1. *If R is a principal ideal domain then the equation $ax + by = 1$ always has a solution for some $x, y \in R$.*

2. *Assume that $ax' + by' = 1$. Then the set of all solutions of $ax + by = 1$ is given by $x = x' + br$, $y = y' - ar$, where r is an arbitrary element of R.*

Proof 1. The set I of elements of S defined by $I \triangleq \{ax + by : x, y \in S\}$ is an ideal of S. Since a and b are coprime and because R is a principal ideal domain, this implies that $I = R$. In particular there exists $x, y \in S$ such that $ax + by = 1$. This proves the first point.
2. We first show that $x = x' + rb$, $y = y' - ra$ are solutions of $ax' + by' = 1$. Indeed

$$ax + by = a(x' + rb) + b(y' - ra) = ax' + rab + by' - rab = ax' + by' = 1.$$

It remains to prove that any solution of $ax + by = 1$ is of the form $x = x' + rb$ and $y = y' - ra$ for some $r \in R$.
Suppose therefore that $ax + by = 1$. Since $ax' + by' = 1$ we have

$$a(x - x') + b(y - y') = 0$$

or

$$a(x - x') = -b(y - y').$$

Since a and b are coprime this implies that a divides $(y - y')$ i.e. $y - y' = -ar$ for some $r \in R$. But then $y = y' - ar$ and it follows from $ax + by = 1$ that $x = x' + br$. The theorem is proved. ■

The parametrization of all the solutions of a Diophantine equation of the form $ax + by = 1$ has been known for a long time for the ring of integers $\mathbf{Z}$. The above theorem is only a restatement for the case of a general ring.

The ring S is a principal ideal domain and the theorem above may thus be used for S.

Theorem 3.3 *Assume that $p \in \mathbf{R}(s)$ and let $c' \in \mathbf{R}(s)$ be a stabilizing controller of p. Let $\frac{n_p}{d_p}$ and $\frac{n'_c}{d'_c}$ be fractional factorizations of p and c' in S. Then the set of all stabilizing controllers of p is given by*

$$Stab(p) = \{\frac{n'_c + rd_p}{d'_c - rn_p} : r \in S\}$$

Proof Since c' is a stabilizing controller, we have

$$n_p n'_c + d_p d'_c = u \in U$$

and thus

$$n_p n'_c u^{-1} + d_p d'_c u^{-1} = 1.$$

By Theorem 3.2, the set of all solutions of $n_p n_c + d_p d_c = 1$ is given by

$$n_c = n'_c u^{-1} + r' d_p u^{-1}$$

and

$$d_c = d'_c u^{-1} - r' n_p u^{-1}.$$

Therefore

$$Stab(p) = \{\frac{n'_c u^{-1} + r' d_p}{d'_c u^{-1} - r' n_p} : r' \in S\}.$$

Multiplying the numerator and denominator of this expression by u and renaming $r = r'u$, we get

$$Stab(p) = \{\frac{n'_c + r d_p}{d'_c - r n_p} : r \in S\}$$

as requested.   ∎

Assume for example that $p = \frac{s}{s-1}$. A fractional factorization of p in S is given by $n_p = \frac{s}{s+1}$ and $d_p = \frac{s-1}{s+1}$. A stabilizing controller for p is $c = -2$. For this controller we may take the trivial factorization $n_c = -2$ and $d_c = 1$ and thus, the set of all stabilizing controllers of p is given by

$$Stab(p) = \{-\frac{2 - r\frac{s-1}{s+1}}{1 - r\frac{s}{s+1}} : r \in S\}.$$

Chosing $r = 1, r = 2$ and $r = \frac{2s}{s+1}$ in the parametrization we get for example the stabilizing controllers $c = s+3$, $c = \frac{4}{s-1}$ and $c = \frac{2(3s+1)}{s-2s-1}$. Note that we obtain the non proper controller $c = s + 3$ for $r = 1$. This point is further analysed in the next section.

The parametrization provides, for a given choice of a stabilizing controller and of one of its factorization n'_c, d'_c, the set of all stabilizing controllers of a system p in terms of a free parameter $r \in S$. To each $r \in S$ corresponds a stabilizing controller and to each stabilizing controller corresponds a stable rational function $r \in S$. Indeed, if c is obtained from the parametrization with $r = r_1$ and also from $r = r_2$, then $\frac{n'_c + r_1 d_p}{d'_c - r_1 n_p} = \frac{n'_c + r_2 d_p}{d'_c - r_2 n_p}$, or $(n'_c + r_1 d_p)(d'_c - r_2 n_p) = (n'_c + r_2 d_p)(d'_c - r_1 n_p)$ but then also $r_1 d_p d'_c - r_2 n_p n'_c = r_2 d_p d'_c - r_1 n_p n'_c$ and, finally, $r_1 = r_2$. Thus, to each stabilizing controller corresponds a unique $r \in S$. The correspondance between S and $Stab(p)$ is bijective.

A remarkable feature of the parametrization is that it is affine in the free parameter $r \in S$ both in the numerator and in the denominator. It is this particular feature that renders the H_∞ methods possible in system theory, namely, the minimization of a weighted H_∞ norm of the sensitivity transfer function $(1 + p(s)c(s))^{-1}$ over the set of all stabilizing controllers. Starting from a factorized stabilizing controller $c' = \frac{n'_c}{d'_c}$ and using the Youla-Kucera parametrization

$$Stab(p) = \{\frac{n'_c + rd_p}{d'_c - rn_p} : r \in S\}$$

we have the following expressions for the sensitivity function:

$$
\begin{aligned}
(1 + pc)^{-1} &= \frac{d_p d_c}{n_p n_c + d_p d_c} \\
&= u^{-1} d_p d_c \\
&= u^{-1} d_p (d'_c - rn_p) \\
&= u^{-1} d_p d'_c - r u^{-1} d_p n_p \\
&= \alpha + r\beta
\end{aligned}
$$

In this last expression we have defined $\alpha = u^{-1} d_p d'_c$ and $\beta = u^{-1} d_p n_p$. Both α and β are fixed once c' is.

When the controller c ranges over $Stab(p)$ the sensitivity transfer function $(1 + pc)^{-1}$ ranges over $\alpha + r\beta$. This last expression is affine in the free parameter r. A minimization, a maximization, or any other optimization strategy conducted on the sensitivity transfer function over the set of all stabilizing controllers is thus relatively easy to perform.

3.3 Finiteness and properness

For practical purposes it is desirable to have controllers that are both finite and proper. Unfortunately, the controllers obtained from the Youla-Kucera parametrization do not always satisfy these two conditions. Also, in later chapters, we will often come accross infinite or non proper controllers when searching for simultaneous stabilizing controllers.

In this section we first show that systems that are simultaneously stabilizable by the infinite or by a non proper controller are also simultaneously stabilizable by a finite proper controller. Properness and finiteness issues are thus not constraining if we are only concerned with existence questions. Thereafter we explain how to change the parametrization so as to obtain only proper and finite controllers.

We start with the finiteness question.

Theorem 3.4 *If $p_i \in \mathbf{R}(s)$ $(i = 1, ..., k)$ are simultaneously stabilizable by the infinite controller, then they are also simultaneously stabilizable by a finite controller.*

Proof A system is stabilizable by the infinite controller if and only if it is inverstable. Thus p_i $(i = 1, ..., k)$ are all inverstable, but then $\frac{1}{p_i}$ are all stable and by the Maximum Modulus Theorem

$$\delta_i = \sup_{s \in \mathbf{C}_{+\infty}} |\frac{1}{p_i(s)}|$$

are all finite numbers. Consider then any $\delta \in \mathbf{R}$ with $0 < \delta < \min_{i=1,\dots,k} \delta_i$ and check that $c = -\frac{1}{\delta}$ is a finite simultaneously stabilizing controller of p_i $(i = 1, \dots, k)$. $\blacksquare$

Next, we consider the properness condition for simultaneous stabilization. Proper controllers have no poles at infinity. A stabilizing controller, if one exists, can always be chosen with no poles at infinity. This result is a corollary of the next more general theorem.

Theorem 3.5 *Consider any two finite sets $Y = \{y_i \in \mathbf{C}_{+\infty} : i = 1, \dots, n\}$ and $Z = \{z_j \in \mathbf{C}_{+\infty} : j = 1, \dots, m\}$ of points in $\mathbf{C}_{+\infty}$. If the k systems $p_i \in \mathbf{R}(s)$ $(i = 1, \dots, k)$ are simultaneously stabilizable, then they are also simultaneously stabilizable with a controller that has no poles in Y and no zeros in Z.*

Proof Factorize the k systems p_i in S by $p_i = \frac{n_i}{d_i}$ and consider a simultaneously stabilizing controller c of factorization $\frac{n_c}{d_c}$ in S. Order the elements of the sets Y and Z so that

$$\alpha_i = |n_c(y_i)| > 0 \ (i = 1, \dots, r) \qquad n_c(y_i) = 0 \ (i = r+1, \dots, n)$$

and

$$\beta_j = |d_c(z_j)| > 0 \ (i = 1, \dots, l) \qquad d_c(z_j) = 0 \ (j = l+1, \dots, m).$$

The controller c is stabilizing and so

$$n_i n_c + d_i d_c = u_i \in U \ (i = 1, \dots, k).$$

Define then

$$\gamma_i = \frac{\inf_{s \in \mathbf{C}_{+\infty}} |u_i(s)|}{\sup_{s \in \mathbf{C}_{+\infty}} |n_t(s) + d_t(s)|} \ (t = 1, \dots, k).$$

Finally, chose any strictly positive number ϵ such that

$$\epsilon < \min\{\min_{i=1,\ldots,r} \alpha_i, \min_{i=1,\ldots,l} \beta_i, \min_{i=1,\ldots,k} \gamma_i\}.$$

Then

$$n_i(n_c + \epsilon) + d_i(d_c + \epsilon) = u_i + \epsilon(n_i + d_i) \in U.$$

Because of our choice of ϵ, the controller defined by

$$c' = \frac{n_c + \epsilon}{d_c + \epsilon}$$

is stabilizing for the systems p_i $(i = 1, ..., k)$.

In addition, c' has neither poles in Y nor zeros in Z. Hence the result. ∎

A corollary of this result is:

Corollary 3.1 *If k systems $p_i \in \mathbf{R}(s)$ $(i = 1, ..., k)$ are simultaneously stabilizable, then they are also simultaneously stabilizable with a proper controller.*

Proof Consider $P = \{\infty\}$ in Theorem 3.5. ∎

As shown by the next two examples, the Youla-Kucera parametrization does not only lead to finite and proper controllers.

Properness

Let $p = \frac{s}{s-1}$. We have shown that the set of all stabilizing controllers of p is given by

$$Stab(p) = \{-\frac{2 - r\frac{s-1}{s+1}}{1 - r\frac{s}{s+1}} : r \in S\}.$$

Choosing $r = 1$ in this expression we get the stabilizing controller

$$c = -\frac{2 - \frac{s-1}{s+1}}{1 - \frac{s}{s+1}} = s + 3,$$

a non proper controller. There exist many other r's that lead to non proper controllers. For example, $r = \frac{s-1}{s+1}$ and $r = \frac{s}{s+2}$ lead, respectively, to

$$\frac{-2 + \frac{s-1}{s+1}\frac{s-1}{s+1}}{1 - \frac{s-1}{s+1}\frac{s}{s+1}} = \frac{-2(s+1)^2 + (s-1)^2}{(s+1)^2 - (s-1)s} = \frac{-s^2 + 6s + 1}{3s + 1},$$

and

$$\frac{-2 + \frac{s}{s+1}\frac{s-1}{s+1}}{1 - \frac{s}{s+2}\frac{s}{s+1}} = \frac{-2(s+2)(s+1)s^2 - s}{(s+2)(s+1) - s^2} = \frac{-s^2 + 7s + 4}{3s + 2}.$$

Both are non-proper controllers.

Finiteness

Let $p = \frac{s+1}{s-1}$. p is inverstable and by Theorem 2.4 a fractional factorization of p is given by $n_p = 1$ and $d_p = \frac{s-1}{s+1}$. A stabilizing controller of p is $c' = \frac{2}{s+1}$ for which we have a fractional factorization of the form $n'_c = \frac{2}{s+1}$ and $d'_c = 1$. The set of all stabilizing controllers of p is given by

$$Stab(p) = \{\frac{\frac{2}{s+1} + r\frac{s-1}{s+1}}{1 - r} : r \in S\}.$$

If $r = 1 \in S$ we obtain the infinite stabilizing controller.

There exist modified parametrizations that allow one to take the properness and finiteness issues into account.

Definition 3.1 *Define the set of finite stabilizing controllers by*

$$Stab_f(p) = \{c \in \mathbf{R}(s) : c \text{ stabilizes } p\}$$

and the set of proper stabilizing controllers by

$$Stab_p(p) = \{c \in \mathbf{R}(s) : c \text{ stabilizes } p : c \text{ is proper}\}.$$

Theorem 3.6 *Assume that $p \in \mathbf{R}(s)$, let $c' \in \mathbf{R}(s)$ be a stabilizing controller of p and let $\frac{n_p}{d_p}$ and $\frac{n'_c}{d'_c}$ be fractional factorizations of p and c' in S. If p is not inverstable then*

$$Stab_f(p) = \{\frac{n'_c + r d_p}{d'_c - r n_p} : r \in S\}.$$

If p is inverstable then

$$Stab_f(p) = \{\frac{n'_c + r d_p}{d'_c - r n_p} : r \in S \setminus \{\frac{d'_c}{n_p}\}\}.$$

Proof The set of all stabilizing controllers of p is given by

$$Stab(p) = \{\frac{n'_c + r d_p}{d'_c - r n_p} : r \in S\}$$

from which we have to excude the situation where $d'_c - r n_p \equiv 0$ if we want to obtain a finite controller. The case $d'_c - r n_p \equiv 0$ occurs only when $d'_c = r n_p$ for some $r \in S$. This situation happens when $r = \frac{d'_c}{n_c}$. But since r also belongs to S, the value $\frac{d'_c}{n_c}$ is to be excluded from the range of r only when n_c divides d'_c. c' is a stabilizing controller of p and thus n_c and d'_c have no common unstable zeros. Therefore n_c divides d'_c if and only if n_c is invertible. That is, if and only if p is inverstable. Hence the result. ∎

In particular, any strictly proper system has an unstable zero (at infinity) and thus the constraint $d'_c - r n_p \not\equiv 0$ is never effective for strictly proper systems.

Theorem 3.7 *Assume that $p \in \mathbf{R}(s)$, let c' be a stabilizing controller of p and let $\frac{n_p}{d_p}$ and $\frac{n'_c}{d'_c}$ be fractional factorizations of p and c' in S. If p is strictly proper then*

$$Stab_p(p) = \{\frac{n'_c + rd_p}{d'_c - rn_p} : r \in S\}.$$

If p is not strictly proper then

$$Stab_p(p) = \{\frac{n'_c + rd_p}{d'_c - rn_p} : r \in S : r(\infty) \neq \frac{d'_c}{n_p}(\infty)\}.$$

Proof Let $p \in \mathbf{R}(s)$ and let c' stabilize p. Let also $\frac{n_p}{d_p}$ and $\frac{n'_c}{d'_c}$ be fractional factorizations of p and c' in S. The set of all stabilizing controllers is given by

$$Stab(p) = \{\frac{n'_c + rd_p}{d'_c - rn_p} : r \in S\}.$$

A controller in this set is proper if and only if it has no pole at infinity. Or, alternatively, if and only if the denominator of any of its fractional factorizations has no zero at infinity.

A denominator of a fractional factorization of c is given by $d'_c - rn_p$ and the controller is thus proper if and only if $d'_c(\infty) - r(\infty)n_p(\infty) \neq 0$.

This condition is automatically satisfied when $n_p(\infty) = 0$, i.e. when p is strictly proper. Indeed, since $n_p n'_c + d_p d'_c \in U$, then, because $n'_c(\infty)$ is finite, $d_p(\infty)d'_c(\infty) \neq 0$ and hence $d'_c(\infty) \neq 0$ which proves that $(d'_c - rn_p)(\infty) \neq 0$ when $n_p(\infty) = 0$.

Suppose now that $n_p(\infty) \neq 0$, meaning that the system is not strictly proper, and choose any r such that $r(\infty) = \frac{d'_c(\infty)}{n_p(\infty)}$. Then the resulting controller is not proper since $(d'_c - rn_p)(\infty) = 0$. ∎

Any $r \in S$ that is such that $r(\infty) \neq \frac{d'_c}{n_p}(\infty)$ leads to a proper controller and since such an r always exists (take, for example, a constant different from the real number $\frac{d'_c}{n'_p}(\infty))$, a proper stabilizing controller always exists.

3.4 Equivalences

Throughout this section we consider k systems $p_i \in \mathbf{R}'(s)$ ($i = 1, \ldots, k$) and we let $\frac{n_i}{d_i}$ be fractional factorizations of p_i in S. We also let $x, y \in S$ be solutions of the equation $n_1 x + d_1 y = 1$ and define

$$a_{ij} = n_i d_j - n_j d_i \ (i, j = 1, \ldots, k)$$

and

$$b_i = n_i x + d_i y \ (i = 1, \ldots, k).$$

The next lemma is needed at several places in the section.

Lemma 3.1 *With the above definitions we have*

$$a_{ij} b_l + a_{jl} b_i + a_{li} b_j = 0 \ (i, j, l = 1, \ldots, k).$$

Proof

$$
\begin{aligned}
a_{ij} b_l + a_{jl} b_i &= (n_i d_j - n_j d_i)(n_l x + d_l y) \\
&\quad + (n_j d_l - n_l d_j)(n_i x + d_i y), \\
&= n_i n_l d_j x + n_i d_j d_l y - n_j n_l d_i x - n_j d_i d_l y \\
&\quad + n_i n_j d_l x + n_j d_i d_l y - n_i n_l d_j x - n_l d_i d_j y, \\
&= n_j x (n_i d_l - n_l d_i) + d_j y (n_i d_l - n_l n_i d_i) \\
&= (n_j x + d_j y)(n_i d_l - n_l d_i) = -b_j a_{li}.
\end{aligned}
$$

■

The first equivalence is between simultaneous and strong stabilization, the second equivalence is between simultaneous and bistable stabilization.

3.4.1 Simultaneous and strong stabilization

The first equivalence is between simultaneous stabilization of k systems and simultaneous stabilization of $k - 1$ systems with a stable controller.

Theorem 3.8 *Assume that $p_i \in \mathbf{R}(s)$ $(i = 1, \ldots, k)$ and let a_{ij} and b_i be defined as above. The k systems p_i are simultaneously stabilizable if and only if the $k - 1$ systems defined by $p_i' \triangleq \frac{a_{i1}}{b_i}$ $(i = 2, \ldots, k)$ are simultaneously stabilizable by a stable controller.*

Proof By Theorem 2.8 there exists a stabilizing controller for p_i if and only if there exist n_c, $d_c \in S$ that simultaneously satisfy

$$n_1 n_c \ + \ d_1 d_c = 1,$$
$$n_i n_c \ + \ d_i d_c \in U \ (i = 2, \ldots, k).$$

A parametrization of all the solutions of the first of these equations is given by $n_c = x + rd_1$ and $d_c = y - rn_1$.
Substituting this parametrization in the second series of equations we get

$$
\begin{aligned}
n_i n_c + d_i d_c &= n_i(x + rd_1) + d_i(y - rn_1), \\
&= n_i x + d_i y + r(n_i d_1 - n_1 d_i), \\
&= b_i + ra_{i1}.
\end{aligned}
$$

The solvability of the original set of simultaneous equations thus depends upon the solvability of the simultaneous equations

$$b_i + ra_{i1} \in U \ (i = 2, \ldots, k).$$

But these last conditions are equivalent to that of the existence of a stable stabilizing controller r for the $k-1$ systems defined by $p_i' = \frac{a_{i1}}{b_i}$ $(i = 2, \ldots, k)$, and hence the theorem is proved. ∎

In this theorem the systems p_i' are defined by $p_i' = \frac{a_{i1}}{b_i} = \frac{n_i d_1 - n_1 d_i}{n_i x + d_i y}$ and this definition involves the stable rational functions x and y that, in turn, are solutions of the Bezout equation $n_1 x + d_1 y = 1$. A more suggestive form for p_i' is given as a corollary of the next theorem.

Theorem 3.9 *Assume that $p_i \in \mathbf{R}(s)$ $(i = 1, \ldots, k)$ and suppose that $q \in \mathbf{R}(s)$ is stable. Then the systems p_i are simultaneously stabilizable if and only if the k systems p_i' defined by $p_i' \triangleq p_i - q$ are simultaneously stabilizable.*

Proof Let $\frac{n_i}{d_i}$ be fractional factorizations of p_i in S. The rational function q is stable and so a fractional factorization of q is given by $q = \frac{q}{1}$. The k systems p_i are simultaneously stabilizable if and only if there exists some $n_c, d_c \in S$ such that

$$n_i n_c + d_i d_c = (n_i + d_i q) n_c + d_i (d_c - q n_c) \in U.$$

This last expression is satisfied if and only if the controller defined by $\frac{n_c}{d_c - q n_c}$ stabilizes the systems defined by $p_i' = \frac{n_i + d_i q}{d_i}$. These systems are also equal to $p_i + q$. The result follows. ∎

A corollary of this theorem is given by:

Corollary 3.2 *Assume that $p_i \in \mathbf{R}(s)$ $(i = 1, \ldots, k)$ and suppose that $p_1 \in \mathbf{R}(s)$ is stable. Then the k systems p_i are simultaneously stabilizable if and only if the $k-1$ systems p_i' defined by $p_i' = p_i - p_1$ $(i = 2, \ldots, k)$ are simultaneously stabilizable by a stable controller.*

Proof Consider Theorem 3.9 with $q = p_1$ and use the fact that the stabilizing controllers of the null systems are stable. ∎

The next corollary is also particularly elegant.

Corollary 3.3 *Assume that $p \in \mathbf{R}(s)$ is stable and let $\Delta_p \in \mathbf{R}(s)$. The systems p and $p + \Delta_p$ are simultaneously stabilizable if and only if Δ_p is strongly stabilizable.*

Proof Consider Corollary 3.2 and remember (again!) that a controller c stabilizes the system 0 if and only if c is stable. ∎

3.4.2 Simultaneous and bistable stabilization

Equivalences contained in this section all follow from the next result.

Theorem 3.10 *Consider $k \geq 3$ and assume that $p_i \in \mathbf{R}(s)$ $(i = 1, ..., k)$. Let a_{ij} and b_i be defined as in the introduction of this section. If p_i are simultaneously stabilizable then there exist $u_i \in U$ $(i = 1, ..., k)$ such that*

$$a_{12}u_i + a_{2i}u_1 + a_{i1}u_2 = 0$$

for $i = 3, ..., k$. If p_i nowhere simultaneously intersect in $\mathbf{C}_{+\infty}$, then these conditions are also sufficient for the k systems to be simultaneously stabilizable.

Proof Suppose that the k systems p_i are simultaneously stabilizable. Then, by Theorem 3.8, we know that there exists some $r \in S$ for which

$$b_i + ra_{i1} = u_i \in U \ (i = 2, ..., k).$$

Using these equalities together with those given in Lemma 3.1 we get for $i = 3, \ldots, k$ (and because $a_{12} = -a_{21}$ and $b_1 = 1$)

$$
\begin{aligned}
a_{12}u_i + a_{2i} + a_{i1}u_2 &= a_{12}(b_i + ra_{i1}) + a_{2i} + a_{i1}(b_2 + ra_{21}) \\
&= a_{i1}(a_{12}r + a_{21}r) + a_{12}b_i + a_{2i} + a_{i1}b_2 \\
&= 0.
\end{aligned}
$$

This proves necessity. To prove sufficiency assume that

$$
a_{12}u_i + a_{2i}u_1 + a_{i1}u_2 = 0 \ (i = 3, \ldots, k).
$$

Then also

$$
a_{2i} = -a_{1i}u + a_{12}u_i \ (i = 3, ..., k)
$$

for $u \in U$ defined by $u = -\frac{u_2}{u_1}$ and u_i redefined as $u_i = -\frac{u_i}{u_1}$. By Lemma 3.1 we also know that since $b_1 = 1$

$$
a_{2i} = -a_{i1}b_2 - a_{12}b_i \ (i = 3, \ldots, k).
$$

Equating these two different expressions of a_{2i} we get

$$
a_{1i}u - a_{12}u_i = a_{i1}b_2 + a_{12}b_i \ (i = 3, \ldots, k)
$$

or

$$
a_{1i}(b_2 + u) = a_{12}(b_i + u_i) \ (i = 3, \ldots, k).
$$

By assumption the systems nowhere simultaneously intersect in $\mathbf{C}_\infty$. If $a_{12}(s_0) = 0$ for $s_0 \in \mathbf{C}_{+\infty}$ then there exists some $i \neq 2$ for which $a_{1i}(s_0) \neq 0$. This implies, because $a_{1i}(b_2 + u) = a_{12}(b_i + u_i)$, that each unstable zero of a_{12} is also an unstable zero of $b_2 + u$. By Theorem 2.2 this means that

$$
b_2 + u = ra_{12} \text{ for some } r \in S.
$$

But then also $a_{1i}r = b_i + u_i$ $(i = 3, \ldots, k)$. And this can be written as

$$a_{12}r - b_2 = u \in U$$

and

$$a_{1i}r - b_i = u_i \in U.$$

Or, defining $r' = -r$

$$a_{1i}r' + b_i \in U \ (i = 2, \ldots, k).$$

But then, applying Theorem 3.8 again, the proof is complete. ∎

Consequences of this theorem are numerous and we give three of those hereafter. The first consequence is an elegant formulation of simultaneous stabilization of three systems that will be used in the forthcoming chapter. The other two results are in the form of equivalent formulations.

Theorem 3.11 *Assume that $p_1, p_2, p_3 \in \mathbf{R}(s)$ do not simultaneously intersect in $\mathbf{C}_{+\infty}$ and let $\frac{n_i}{d_i}$ be fractional factorizations of p_i in S. These three systems are simultaneously stabilizable if and only if there exist u_1, u_2 and $u_3 \in U$ such that*

$$(n_1 d_2 - n_2 d_1)u_3 + (n_2 d_3 - n_3 d_2)u_1 + (n_3 d_1 - n_1 d_3)u_2 = 0.$$

Proof Apply Theorem 3.10 for $k = 3$. ∎

For the first equivalence we need to introduce the notion of simultaneous partial pole assignment.

Definition 3.2 (Ghosh [55]) *Assume that $p, c \in \mathbf{R}(s)$ and let $\frac{n_p}{d_p}$ and $\frac{n_c}{d_c}$ be fractional factorizations of p and c in S. If $n_p n_c +$*

$d_p d_c \in U$ then c stabilizes p. If $n_p n_c + d_p d_c$ has unstable zeros at $s_0, s_1, ..., s_n \in \mathbf{C}_{+\infty}$, then c assigns the closed loop unstable poles of p at $s_0, s_1, ..., s_n$. These unstable zeros are the unstable poles of the closed loops associated to the system p controlled by c.

Theorem 3.12 *Assume that $p_i \in \mathbf{R}(s)$ $(i = 1, \ldots, k)$ and suppose that p_i do not simultaneously intersect in $\mathbf{C}_{+\infty}$. Then the k systems p_i are simultaneously stabilizable if and only if there exists a bistable controller that assigns the closed loop unstable poles of the $k - 2$ systems defined by $p_i' = \frac{a_{2i}}{a_{i1}}$ $(i = 3, \ldots, k)$ at the unstable zeros of a_{12}.*

Proof By Theorem 3.10 the systems p_i $(i = 1, ..., k)$ are simultaneously stabilizable if and only if there exist $u_i \in U$ $(i = 1, ..., k)$ such that

$$a_{12} u_i + a_{2i} u_1 + a_{i1} u_2 = 0$$

for $i = 3, ..., k$. This last condition is equivalent to

$$a_{2i} u_1 + a_{i1} u_2 = -a_{12} u_i$$

and thus the proof is complete. ∎

A perhaps more suggestive form of the theorem is the following corollary. If p_1 avoids p_2 in $\mathbf{C}_{+\infty}$ then $a_{12} \in U$ and the assignment of the closed loop unstable poles at the unstable zeros of a_{12} is just the same as stabilization. In addition to this, if $a_{12} \in U$, we automatically have that $p_i(s)$ $(i = 1, \ldots, k)$ nowhere simultaneously interest in $\mathbf{C}_{+\infty}$. We have thus:

Corollary 3.4 *Assume that $p_i \in \mathbf{R}(s)$ $(i = 1, \ldots, k)$ and suppose that p_1 avoids p_2 in $\mathbf{C}_{+\infty}$. Then the k systems p_i $(i = 1, \ldots k)$ are* *simultaneously stabilizable if and only if the $k-2$ systems defined by* $p'_i = \frac{a_{2i}}{a_{i1}}$ $(i = 3, \ldots, k)$ *are simultaneously stabilizable by a bistable controller.*

As a final remark let us note that these results are still valid in a general ring context. In particular, they all have counterparts for the general concept of Λ-stability.

3.5 Summary and bibliography

This chapter revolved around the Youla-Kucera parametrization.

The parametrization of all the solutions of a Bezout equation in the ring of integers is an old and well established result (see for example [49]). The same conceptual ideas applied to the ring S lead to the Youla-Kucera parametrization. We have made some comments and derived improvements of this parametrization that allow, for example, to accomodate properness restrictions on the controller.

The parametrization is then used to derive equivalences. Roughly speaking, the equivalences are between simultaneous stabilization of k systems, strong simultaneous stabilization of $k-1$ systems and bistable simultaneous stabilization of $k-2$ systems.

The Youla-Kucera parametrization is to be found in Youla *et al.* [124], Desoer, Liu, Murray and Saeks [33], Vidyasagar and Viswanadham [110] or Kucera [80].

For the first equivalences, see for example Saeks and Murray [102] or Vidyasagar and Viswanadham [110].

A weaker form of the second equivalences is given in Ghosh [50,53, 54]. See also Wei for this [119].

Chapter 4

Necessary conditions: interlacement

No dimensions pointland, one dimension lineland,
two dimensions flatland,... three dimensions spaceland;
a romance of many dimensions.

E. Abbot, Flatland.

4.1 Introduction

When not referring explicitly to any instability region we mean $C_{+\infty}$-stability.

A controller stabilizes a system if and only if the corresponding four closed loop transfer functions $pc(1 + pc)^{-1}$, $p(1 + pc)^{-1}$, $c(1 + pc)^{-1}$ and $(1+pc)^{-1}$ have no poles in the extended right half plane $C_{+\infty}$. A weaker requirement would be that the closed loop transfer functions have no poles on the extended positive real line $\mathbf{R}_{+\infty}$. According to our definitions, a controller that satisfies such a weakened condition is $\mathbf{R}_{+\infty}$-stabilizing.

Systems that are simultaneously stabilizable (i.e. simultaneously $\mathbf{C}_{+\infty}$-stabilizable) are also simultaneously $\mathbf{R}_{+\infty}$-stabilizable. Conditions for simultaneous $\mathbf{R}_{+\infty}$-stabilization are thus necessary conditions for simultaneous stabilization. In this chapter we study these conditions. The central theme of the chapter is: under what condition does there exist a controller that simultaneously $\mathbf{R}_{+\infty}$-stabilizes k given systems? These conditions encompass all the presently available necessary conditions for simultaneous stabilization.

The question of simultaneous stabilization may be seen as a question of avoidance in $\mathbf{C}_{+\infty}$. We have also shown that the ideas associated to $\mathbf{C}_{+\infty}$-stabilization can be extended to Λ-stabilization where Λ is an arbitrary subset of $\mathbf{C}_{\infty}$. The problem of simultaneous $\mathbf{R}_{+\infty}$-stabilization can thus be seen as one of avoidance on $\mathbf{R}_{+\infty}$.

One of the drawbacks of the avoidance interpretation is that it is usually hard to see –to represent mentally– where two rational functions intersect in the complex plane. This is because, in general, rational functions map complex numbers to complex numbers and, hence, four dimensions are needed to represent their behaviour. The situation is different when we consider the real line only. On the real line, a real rational function takes real values only and it is then relatively easy to see if, and where, two rational functions intersect on the real axis. This fact coupled with the link between stabilization and avoidance gives powerful insights into $\mathbf{R}_{+\infty}$-stabilization.

It is this idea that allows us to give a geometrical treatment of simultaneous $\mathbf{R}_{+\infty}$-stabilization: k systems $p_i \in \mathbf{R}(s)$ $(i = 1, ..., k)$ are simultaneouly $\mathbf{R}_{+\infty}$-stabilizable if and only if there exists a rational function $q(s)$ such that $q(s) \neq p_i(s)$ for all $s \in \mathbf{R}_{+\infty}$ and $i = 1, ..., k$. The results and theorems of this chapter all start from this idea and rely on simple geometrical arguments.

Section 2 deals with strong $\mathbf{R}_{+\infty}$-stabilization and with simultaneous $\mathbf{R}_{+\infty}$-stabilization of two systems. In that section we introduce the parity interlacing condition.

The last two sections present completely general conditions for simultaneous $\mathbf{R}_{+\infty}$-stabilization of k systems when k is greater than or equal to three. There are two different cases to consider and these are analysed separately in Section 3 and in Section 4.

In Section 3 we deal with the situation where the systems do not simultaneously intersect on the extended positive real line. In that section we introduce the even and the k-interlacing conditions. These conditions may be thought of as extensions of the parity interlacing condition for the case of more than two systems. Bistable stabilization of a single system p is equivalent to simultaneous stabilization of the three systems p, 0 and ∞. 0 never intersects ∞ on $\mathbf{R}_{+\infty}$ and thus the bistable stabilization is also analysed in Section 3.

In Section 4 we deal with the situation where the systems do simultaneously intersect on the extended positive real line. That is, we analyse the case where $p_i(s_0) = p_j(s_0)$ for some $s_0 \in \mathbf{R}_{+\infty}$ and all $i, j = 1, ..., k$. In this situation the result is surprisingly easy to express: the systems are simultaneously $\mathbf{R}_{+\infty}$-stabilizable if and only if they are pairwise simultaneously $\mathbf{R}_{+\infty}$-stabilizable. The proof of this result needs the introduction of the concept of winding transform and of winding number.

4.2 Two systems and strong stabilization

The $\mathbf{R}_{+\infty}$-stabilizability of a system by an $\mathbf{R}_{+\infty}$-stable controller relies only on the pattern of the real positive poles and zeros of the system.

Definition 4.1 *Assume that p is in $\mathbf{R}(s)$. Then p satisfies the parity interlacing condition if and only if p has an even number of poles between each pair of zeros on the extended positive real axis.*

In the definition and as everywhere else in the monograph poles and zeros are counted with multiplicity included: the system $\frac{1}{(s-2)^2}$ has two poles at $s = 2$ and two zeros at infinity.

The succession of poles (P) and zeros (Z) of the system $p_1 = \frac{(s-1)(s-3)}{s(s-2)}$ on $\mathbf{R}_{+\infty}$ is PZPZ whereas that of the system $p_2 = \frac{s}{(s-1)^2}$ is ZPPZ (the last zero is at infinity). The second of these systems satisfies the parity interlacing condition but the first one does not.

Another useful way to express the parity interlacing condition is by means of the non oriented graph G_p (the p stands for *parity*) represented in Figure 4.1. A system p satisfies the parity interlacing condition if and only if the succession of poles and zeros of p, as s increases from 0 to $+\infty$, describes a succession of edges of a path in the graph G_p. ZPPZ corresponds to a succession of edges of a path in the graph G_p whereas PZPZ does not.

Figure 4.1: The graph G_p. A system $p(s)$ satisfies the parity interlacing condition if and only if the succession of poles and zeros of $p(s)$, as s increases from 0 to $+\infty$, describes a succession of edges of a path in G_p.

In the next theorem we prove the equivalence between strong $\mathbf{R}_{+\infty}$-stabilizability and the parity interlacing condition. We therefore need a preliminary lemma.

Lemma 4.1 Let $s_i \in \mathbf{R}_{+\infty}$ and $v_i \in \mathbf{R}$ $(i = 1, ..., n)$. There exists $u \in U(\mathbf{R}_{+\infty})$ such that $u(s_i) = v_i$ $(i = 1, ..., n)$ if and only if the next two conditions are satisfied:

1. $v_i \neq 0$ $(i = 1, ..., n)$.

2. v_i have the same sign.

Proof Necessity is obvious because any $u \in U(\mathbf{R}_{+\infty})$ is continuous, real valued and never equal to zero on $\mathbf{R}_{+\infty}$. Hence $u(s)$ has the same sign everywhere on $\mathbf{R}_{+\infty}$: it is either positive or negative on the positive real axis.

For sufficiency suppose without loss of generality that all the v_i are positive and define

$$v_m = \min_{i=1,\ldots,n} v_i > 0$$

and

$$v_i' = \sqrt{v_i - v_m}.$$

Find a polynomial $p \in \mathbf{R}[s]$ such that

$$p(s_i) = v_i' \ (i = 1, \ldots, n)$$

and define

$$v(s) \triangleq v_m + (p(s))^2.$$

Finally, find any polynomial $w(s)$ that has no zeros on $\mathbf{R}_{+\infty}$, that has same degree as $v(s)$ and that is such that

$$w(s_i) = 1 \ (i = 1, \ldots, n).$$

Define $u(s) \triangleq \frac{v(s)}{w(s)}$ and verify that $u(s)$ satisfies the conditions of the lemma. ∎

The proof of the main theorem of this section is now almost trivial.

Theorem 4.1 *Assume that $p \in \mathbf{R}(s)$. Then the system p is strongly $\mathbf{R}_{+\infty}$-stabilizable if and only if it satisfies the parity interlacing condition.*

Proof Consider any fractional factorization $\frac{n_p}{d_p}$ of p in $S(\mathbf{R}_{+\infty})$. p is strongly $\mathbf{R}_{+\infty}$-stabilizable if and only if there exists some $c \in S(\mathbf{R}_{+\infty})$ such that

$$n_p c + d_p \in U(\mathbf{R}_{+\infty}).$$

This condition is satisfied if and only if there exists a $u \in U(\mathbf{R}_{+\infty})$ such that n_p divides $u - d_p$ in $S(\mathbf{R}_{+\infty})$. By Theorem 2.2 this is possible if and only if there exists a $u \in U(\mathbf{R}_{+\infty})$ with $u(s) = d_p(s)$ for all $s \in \mathbf{R}_{+\infty}$ for which $n_p(s) = 0$. By the previous lemma, a necessary and sufficient condition for this is that $d_p(s)$ has the same sign whenever $n_p(s) = 0$ on $\mathbf{R}_{+\infty}$ or, equivalently, that $d(s)$ has an even number of zeros between each pair of zeros of $n(s)$. ∎

Connected with strong $\mathbf{R}_{+\infty}$-stabilization is simultaneous $\mathbf{R}_{+\infty}$-stabilization of two systems. The next theorem provides a constructive checking procedure for simultaneous $\mathbf{R}_{+\infty}$-stabilization of two systems.

Theorem 4.2 *Assume that p_1 and $p_2 \in \mathbf{R}(s)$ and let $\frac{n_i}{d_i}$ be fractional factorizations of p_i in $S(\mathbf{R}_{+\infty})$ ($i = 1, 2$). Define $x, y \in S(\mathbf{R}_{+\infty})$ by*

$$n_1 x + d_1 y = 1.$$

Then p_i ($i = 1, 2$) are simultaneously $\mathbf{R}_{+\infty}$-stabilizable if and only if the system p defined by

$$p = \frac{n_1 d_2 - n_2 d_1}{n_2 x + d_2 y}$$

satisfies the parity interlacing condition.

Proof By Theorem 3.8 the systems p_1 and p_2 are simultaneously $\mathbf{R}_{+\infty}$-stabilizable if and only if p is strongly $\mathbf{R}_{+\infty}$-stabilizable. This last condition is satisfied if and only if p satisfies the parity interlacing condition, hence the result. ∎

4.3 Non intersecting systems

4.3.1 Introduction

The systems $p_i \in \mathbf{R}(s)$ $(i = 1, ..., k)$ simultaneously intersect on $\mathbf{R}_{+\infty}$ if $p_i(s_0) = p_j(s_0)$ for some $s_0 \in \mathbf{R}_{+\infty}$ and all $i, j = 1, ..., k$.

Simultaneous $\mathbf{R}_{+\infty}$-stabilization conditions for k systems are different depending on whether or not the systems simultaneously intersect on $\mathbf{R}_{+\infty}$.

We have split our analysis in accordance with this observation. This section deals with the case where the systems do not simultaneously intersect on $\mathbf{R}_{+\infty}$. The next section is an analysis of the intersecting case.

The simultaneous $\mathbf{R}_{+\infty}$-stabilization conditions for non intersecting systems are similar to the parity interlacing conditions. In the first subsection we introduce the 3-interlacing condition and we show that three systems are simultaneously $\mathbf{R}_{+\infty}$-stabilizable if and only if they satisfy this condition. Bistable stabilization is a special case of simultaneous stabilization of three non intersecting systems and we perform, for the sake of clarity, a separate study of bistable stabilization in the second subsection. When considering bistable stabilization, the 3-interlacing condition degenerates into the even interlacing condition.

In the last subsection we introduce the k-interlacing condition for k greater than or equal to four and we show that it is a necessary condition for simultaneous $\mathbf{R}_{+\infty}$-stabilization of k systems. A construction of the 4-interlacing condition is also provided.

4.3.2   The 3-interlacing condition

The simultaneous $\mathbf{R}_{+\infty}$-stabilizability of three systems that do not simultaneously intersect depends only on the pairwise intersections of the three systems on $\mathbf{R}_{+\infty}$.

Theorem 4.3 *Let* $p_1, p_2, p_3, p'_1, p'_2, p'_3 \in R(s)$. *Assume that* p_1, p_2 *and* p_3 *do not simultaneously intersect on* $\mathbf{R}_{+\infty}$ *and that the pairwise intersections of* p_i *and of* p'_i *on* $\mathbf{R}_{+\infty}$ *are the same. That is*

$$p_i(s) = p_j(s), \; s \in \mathbf{R}_{+\infty} \Leftrightarrow p'_i(s) = p'_j(s), \; s \in \mathbf{R}_{+\infty}.$$

Then p_1, p_2, p_3 *are simultaneously* $\mathbf{R}_{+\infty}$-*stabilizable if and* p'_1, p'_2, p'_3 *are.*

Proof Let $p_i = \frac{n_i}{d_i}$ $(i = 1, 2, 3)$ and $p'_i = \frac{n'_i}{d'_i}$ $(i = 1, 2, 3)$ be arbitrary fractional factorizations of p_i and p'_i in $S(\mathbf{R}_{+\infty})$. As in Section 3.4, define $a_{ij} = n_i d_j - n_j d_i$ $(i, j = 1, 2, 3)$ and $a'_{ij} = n'_i d'_j - n'_j d'_i$ $(i, j = 1, 2, 3)$. By the version of Theorem 3.10 for $\mathbf{R}_{+\infty}$-stability, the systems p_i are simultaneously $\mathbf{R}_{+\infty}$-stabilizable if and only if there exist u_1, u_2 and $u_3 \in U(\mathbf{R}_{+\infty})$ such that

$$a_{12}u_3 + a_{23}u_1 + a_{31}u_2 = 0.$$

The pairwise intersections of p_i and of p'_i on $\mathbf{R}_{+\infty}$ are the same and thus p'_i do not simultaneously intersect. Therefore, by the same theorem, p'_i are simultaneously $\mathbf{R}_{+\infty}$-stabilizable if and only if there exist u'_1, u'_2 and $u'_3 \in U(\mathbf{R}_{+\infty})$ such that

$$a'_{12}u_3 + a'_{23}u_1 + a'_{31}u_2 = 0.$$

By assumption, the pairwise intersections between p_i and p'_i in $\mathbf{R}_{+\infty}$ are the same and thus

$$a_{12} = a'_{12}v_3$$

$$a_{23} = a'_{23}v_1$$

and

$$a_{31} = a'_{31}v_2$$

for some v_1, v_2 and $v_3 \in U(\mathbf{R}_{+\infty})$. Defining $u'_{12} = u_{12}v_3$, $u'_{23} = u_{23}v_1$ and $u'_{31} = u_{31}v_2$, the theorem is proved. ∎

Thus the stabilizability of three systems that do not simultaneously intersect depends only on the pairwise intersections. The systems themselves do not really matter, it is their pairwise intersections that renders stabilization possible or not. A description of the triplets of systems that are simultaneously stabilizable is as follows.

Theorem 4.4 *Assume that the three systems p_1, p_2 and $p_3 \in \mathbf{R}(s)$ do not simultaneously intersect on $\mathbf{R}_{+\infty}$. Then p_1, p_2 and p_3 are simultaneously $\mathbf{R}_{+\infty}$-stabilizable if and only if there exist three $\mathbf{R}_{+\infty}$-stable systems $p'_i \in \mathbf{R}(s)$ that have the same pairwise intersections on $\mathbf{R}_{+\infty}$ as p_i.*

Proof For sufficiency note that $\mathbf{R}_{+\infty}$-stable systems are always simultaneously $\mathbf{R}_{+\infty}$-stabilized by the controller 0. If there exist three $\mathbf{R}_{+\infty}$-stable systems $p'_i \in \mathbf{R}(s)$ that have the same pairwise intersections in $\mathbf{R}_{+\infty}$ as p_i, then p'_i are simultaneously $\mathbf{R}_{+\infty}$-stabilizable and an application of the above Theorem 4.3 shows that p_i are also simultaneously $\mathbf{R}_{+\infty}$-stabilizable.

To prove necessity, suppose that p_1, p_2 and p_3 are simultaneously stabilizable and that they do not simultaneously intersect on $\mathbf{R}_{+\infty}$.

Let $p_i = \frac{n_i}{d_i}$ $(i = 1, 2, 3)$ be arbitrary fractional factorizations in $S(\mathbf{R}_{+\infty})$ and define $a_{ij} = n_i d_j - n_j d_i$ $(i, j = 1, 2, 3)$. By Theorem 3.10 there exist $u_1, u_2, u_3 \in U(\mathbf{R}_{+\infty})$ such that

$$a_{12}u_3 + a_{23}u_1 + a_{31}u_2 = 0.$$

Consider any $p_2' \in S(\mathbf{R}_{+\infty})$ and define $p_1' = p_2' + u_3 a_{12} \in S(\mathbf{R}_{+\infty})$ and $p_3' = p_2' - u_1 a_{23} \in S(\mathbf{R}_{+\infty})$. Then we have $r_1 - r_2 = a_{12}u_3$, $r_2 - r_3 = a_{23}u_1$, but also $p_3' - p_1' = a_{31}u_2$. And thus p_1', p_2' and p_3' are $\mathbf{R}_{+\infty}$-stable and have pairwise the same intersections in $\mathbf{R}_{+\infty}$ as p_i. This ends the proof. ∎

In this proof we use only algebraic properties of $S(\mathbf{R}_{+\infty})$ and, because these properties are shared with $S(\Lambda)$ for any $\Lambda \subset \mathbf{C}_\infty$, the proof remains valid for any instability zone Λ. If we modify three systems into three other systems in such a way that the pairwise intersections in the instability zone are left unchanged then so is their stabilizability.

Theorem 4.5 *Assume that $\Lambda \subset \mathbf{C}_\infty$. Three systems that do not simultaneously intersect in Λ are simultaneously Λ-stabilizable if and only if there exist three Λ-stable systems that have pairwise the same intersections in Λ as the three original systems.*

We illustrate this theorem with an example in a $\mathbf{C}_{+\infty}$-stability set-up. Let $p_1 = -\frac{1}{2}$, $p_2 = \frac{1}{s}$ and $p_3 = \frac{3}{(s-1)}$. These three systems nowhere simultaneously intersect in $\mathbf{C}_{+\infty}$. They have pairwise the same intersections in $\mathbf{C}_{+\infty}$ as $p_1' = 2, p_2' = \frac{s}{s+1}$ and $p_3' = \frac{s-1}{s+2}$. The latter three systems are all stable and, hence, are simultaneously stabilizable by a constant controller (in particular, they are stabilized by $c \equiv \infty$). Thus p_i $(i = 1, 2, 3)$ are simultaneously stabilizable.

When the instability zone is the real positive axis one can find a tractable test to decide whether there exist functions in $S(\mathbf{R}_{+\infty})$ that have given sets of pairwise intersections. This test is the 3-interlacing condition.

Definition 4.2 *The graph G_3 represented in Figure 4.2 (a) (see page 82) is constructed in the following way: there are 6 vertices labelled by the permutations $\underline{i}$ of $(1, 2, 3)$ ($\underline{i} = (i_1, i_2, i_3)$ where $i_\alpha \in \{1, 2, 3\}$ and $i_\alpha \neq i_\beta$ for $\alpha \neq \beta$). There is an edge between the vertices $\underline{i}$ and $\underline{j}$ if and only if their labels differ only by the transposition of two successive coordinates ($i_\alpha = j_{\alpha+1}$ and $i_{\alpha+1} = j_\alpha$ for $\alpha = 1$ or $\alpha = 2$ and $i_\beta = j_\beta$ for $\beta \neq \alpha$). The edge is then labelled by (i_α, j_α). Assume that $p_1, p_2, p_3 \in \mathbf{R}(s)$ and that p_i do not simultaneously intersect on $\mathbf{R}_{+\infty}$. p_1, p_2 and p_3 satisfy the 3-interlacing condition if and only if the succession of their intersections, as s increases from 0 to $+\infty$, describes a succession of edges of a path in the graph G_3.*

The central result of this section, and the reason for the introduction of the 3-interlacing condition, is the following theorem.

Theorem 4.6 *Assume that $p_1, p_2, p_3 \in \mathbf{R}(s)$ do not simultaneously intersect on $\mathbf{R}_{+\infty}$. p_i are simultaneously $\mathbf{R}_{+\infty}$-stabilizable if and only if they satisfy the 3-interlacing condition.*

Proof Due to Theorem 4.4, the three systems p_i are simultaneously $\mathbf{R}_{+\infty}$-stabilizable if and only if there exist three $\mathbf{R}_{+\infty}$-stable rational functions p'_i that have on $\mathbf{R}_{+\infty}$ the same pairwise intersections as p_i. It remains to show that this last condition is equivalent to the 3-interlacing condition. That is, we have to show that

1. the succession of intersections of three $\mathbf{R}_{+\infty}$-stable rational functions, as s increases from 0 to infinity, describes a succession of edges of a path in the graph G_3,

2. from any given sets of intersections whose succession describes a path in the graph G_3 it is possible to construct a corresponding set of three $\mathbf{R}_{+\infty}$-stable rational functions that have, pairwise, these sets for sets of intersection.

For point 1, assume that p_1, p_2 and $p_3 \in \mathbf{R}(s)$ have no poles on $\mathbf{R}_{+\infty}$ and suppose that we have, for example, the inequalities

$$p_1(s_0) > p_2(s_0) > p_3(s_0)$$

for some $s_0 \in \mathbf{R}_{+\infty}$. Denote by s_1 the smallest next point of pairwise intersection that is larger than s_0 and construct s_2 similarly. At any point $s \in [s_1, s_2]$ it is clear that, due to the continuity and differentiability of the functions p_i, there is a reversal of one, and of one only, of the strict inequalities above. For example, if

$$p_1(s_0) > p_2(s_0) > p_3(s_0)$$

and $s \in [s_1, s_2]$, then either

$$p_2(s) > p_1(s) > p_3(s)$$

or

$$p_1(s) > p_3(s) > p_2(s).$$

The intersections at $s = s_1$ corresponding to these two cases are intersections between p_1 and p_2 or between p_2 and p_3. These possible reversals are exactly those that are formalized in the graph G_3 by the edges that start from the vertex $(1, 2, 3)$ and that go either to the vertex $(2, 1, 3)$ or to the vertex $(1, 3, 2)$. These two vertices are labelled, respectively, by $(1,2)$ and by $(2,3)$. From this it is clear that the sequence of pairwise intersections, as s increases from 0 to $+\infty$, describes a path in the graph G_3.

From the construction of the graph it is also clear that any set of pairwise intersections whose succession fits in the graph G_3 can be obtained as intersections of $\mathbf{R}_{+\infty}$-stable rational functions. ∎

To illustrate the use of the theorem we analyse an example of three systems that are pairwise simultaneously stabilizable but that are not simultaneously $\mathbf{R}_{+\infty}$-stabilizable (the example is taken from Ghosh [52]).

The systems are $p_1 = \frac{s-7}{s-4.6}$, $p_2 = \frac{s-2}{2s-2.6}$ and $p_3 = \frac{s-6}{4.9s-24.6}$. The intersections between p_1 and p_2 are at $\sigma_{12} = 1$ and $\sigma_{12} = 9$. For the other two pairwise intersections, we get $\sigma_{23} = 3$ and $\sigma_{23} = 4$, $\sigma_{31} = 7.34$ and $\sigma_{31} = 5.17$. All the intersections happen to be on $\mathbf{R}_{+\infty}$ but this is by no way generic. Ordering the succession of the pairwise intersections on $\mathbf{R}_{+\infty}$, we get, $12, 23, 23, 31, 31, 12$. This does not correspond to a succession of edges of a path in the graph G_3. Hence p_i $(i = 1, 2, 3)$ do not satisfy the 3-interlacing condition and, by Theorem 4.6, p_1, p_3 and p_3 are not simultaneously $\mathbf{R}_{+\infty}$-stabilizable.

As a final remark note that the 3-interlacing condition can be expressed in a more compact form. Due to the symmetry in the graph G_3 any succession of labels obtained from a path in graph G_3 is also obtainable from a path in the reduced version G_3' and vice-versa (see Figure 4.2 (b), page 82). The two graphs G_3 and G_3' are equivalent in terms of the succession of edges.

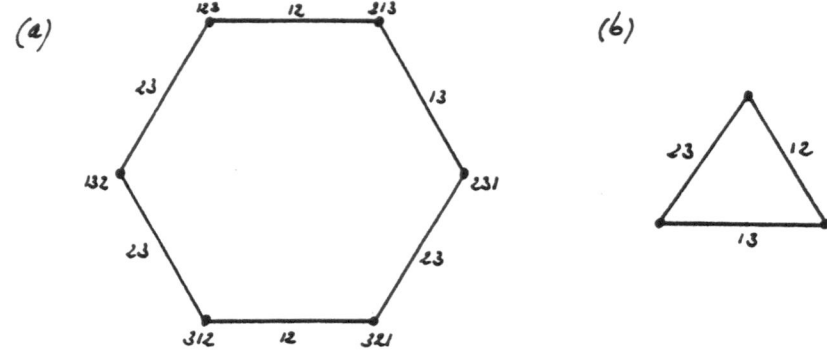

Figure 4.2: (a) The graph G_3 for the 3-interlacing property and (b) its reduced version G_3'. The systems $p_1(s), p_2(s)$ and $p_3(s)$ satisfy the 3-interlacing condition if and only if the succession of their intersections, as s increases from 0 to $+\infty$, describes a succession of edges of a path in G_3 or in G_3'.

4.3.3 The even interlacing condition

A system p that is bistably $\mathbf{R}_{+\infty}$-stabilizable is, by definition, $\mathbf{R}_{+\infty}$-stabilizable by an $\mathbf{R}_{+\infty}$-stable controller and by an $\mathbf{R}_{+\infty}$-inverstable controller. A necessary condition for $\mathbf{R}_{+\infty}$-stabilizability by an $\mathbf{R}_{+\infty}$-stable controller is the parity interlacing property and a similar condition for the $\mathbf{R}_{+\infty}$-stabilizability by an $\mathbf{R}_{+\infty}$-inverstable controller is that p^{-1} satisfies the parity interlacing property. From this it is clear that a system that is bistably $\mathbf{R}_{+\infty}$-stabilizable satisfies the parity interlacing condition and its inverse formulation. If we merge these two conditions in a single one we get the next definition.

Definition 4.3 *A system satisfies the even interlacing condition if and only if it has an even number of poles between each pair of zeros and an even number of zeros between each pair of poles on the extended positive real axis.*

The system $p_1 = \frac{(s-1)(s-3)}{(s-2)(s-4)}$ does not satisfy the even interlacing property since the succession of poles and zeros of p_1 is ZPZP. The system $p_2 = \frac{(s-1)(s-3)^2}{(s-2)(s-2.5)(s-4)}$ satisfies the even interlacing property.

Like the parity interlacing property, the even interlacing property can be expressed in a condensed way by a graph. See the graph G_e (the e stands for *even*) in Figure 4.3, page 84. A system $p \in \mathbf{R}(s)$ satisfies the even interlacing condition if and only if the succession of its poles and zeros on the extended positive real axis describes a succession of edges of a path in G_e. The even interlacing condition is a necessary condition for bistable $\mathbf{R}_{+\infty}$-stabilization. It is also sufficient because bistable $\mathbf{R}_{+\infty}$-stabilization of p is equivalent to simultaneous $\mathbf{R}_{+\infty}$-stabilization of the three systems p, 0 and ∞ and the 3-interlacing condition degenerates into the even interlacing condition in this case.

Theorem 4.7 (Wei [118]) *The system $p \in \mathbf{R}(s)$ is bistably $\mathbf{R}_{+\infty}$-stabilizable if and only if it satisfies the even interlacing condition.*

Proof The system p is bistably $\mathbf{R}_{+\infty}$-stabilizable if and only if the three systems $p_1 = p$, $p_2 = 0$ and $p_3 = \infty$ are simultaneously $\mathbf{R}_{+\infty}$-stabilizable. There are no intersections between p_2 and p_3 on $\mathbf{R}_{+\infty}$, the intersections between p_1 and p_2 on $\mathbf{R}_{+\infty}$ are the zeros of p on $\mathbf{R}_{+\infty}$ and the intersections between p_1 and p_3 on $\mathbf{R}_{+\infty}$ are the poles of p on $\mathbf{R}_{+\infty}$. The conditions of Theorem 4.6 are satisfied and the three systems above are simultaneously $\mathbf{R}_{+\infty}$-stabilizable if and only if they satisfy the 3-interlacing property. This condition, expressed for the case above, is equivalent to the even interlacing property, hence the result. ∎

Figure 4.3: The graph G_e for the even interlacing condition. A system $p(s)$ satisfies the even interlacing condition if and only if the succession of poles and zeros of $p(s)$, as s increases from 0 to $+\infty$, describes a succession of edges of a path in G_e.

4.3.4 The k-interlacing condition

The idea of the 3-interlacing condition can be extended to the case of more than 3 systems so as to obtain necessary conditions for simultaneous $\mathbf{R}_{+\infty}$-stabilization of k systems. It is yet unknown if these conditions are sufficient.

Definition 4.4 (k-interlacing condition) *For any integer k ($k \geq 3$) the graph G_k is constructed as follows: there are $k!$ vertices labelled by the permutations $\underline{i}$ of $(1, 2, ..., k)$ ($\underline{i} = (i_1, i_2, ..., i_k)$ with $i_\alpha \in \{1, 2, ..., k\}$ and $i_\alpha \neq i_\beta$ for $\alpha \neq \beta$). There is an edge between the vertices $\underline{i}$ and $\underline{j}$ if and only if these two edges differ only by* *the transposition between two successive coordinates ($i_\alpha = j_{\alpha+1}$ and $i_{\alpha+1} = j_\alpha$ for some $\alpha \in \{1, ..., k-1\}$ and $i_\beta = j_\beta$ for all $\beta \neq \alpha$). The corresponding edge is labelled by (i_α, j_α). The resulting graph obtained for $k = 4$ is represented in Figure 4.4 (a) (page 87) where we have, for the sake of clarity, omitted the labels of the vertices. k systems $p_i \in \mathbf{R}(s)$ that do not intersect three by three on $\mathbf{R}_{+\infty}$ satisfy the k-interlacing condition if and only if the succession of their intersections, as s increases from 0 to infinity, describes a succession of edges of a path in the graph G_k.*

We now prove the result that justifies the introduction of the k-interlacing condition.

Theorem 4.8 *Assume that $p_i \in \mathbf{R}(s)$ ($i = 1, ..., k$) do not simultaneously intersect three by three on the real positive axis. If p_i are simultaneously $\mathbf{R}_{+\infty}$-stabilizable then they satisfy the k-interlacing condition.*

Proof Let $\frac{n_i}{d_i}$ be fractional factorizations of the systems p_i in $S(\mathbf{R}_{+\infty})$ and consider the fractional factorization of a simultaneously stabilizing controller $c = \frac{n_c}{d_c}$. There exists some $x, y \in S(\mathbf{R}_{+\infty})$ such that

$$n_c x + d_c y = 1.$$

Since the k systems p_i are simultaneously $\mathbf{R}_{+\infty}$-stabilized by c we have

$$n_i n_c + d_i d_c = u_i \in U(\mathbf{R}_{+\infty}).$$

We define k $\mathbf{R}_{+\infty}$-stable rational functions p_i' ($i = 1, ..., k$) by

$$p_i' = \frac{n_i y - d_i x}{u_i}.$$

Elementary algebraic manipulations show that, for $i, j = 1, ..., k$ and $s \in \mathbf{R}_{+\infty}$

$$p'_i(s) = p'_j(s) \Leftrightarrow n_i(s)d_j(s) - n_j(s)d_i(s) = 0.$$

That is, $p'_i \in \mathbf{R}(s)$ $(i = 1, ..., k)$ have no poles on $\mathbf{R}_{+\infty}$ and have the same pairwise intersections as p_i on $\mathbf{R}_{+\infty}$. The rest of the proof is similar to the first part of the proof of Theorem 4.6. ■

The 4-interlacing property can be given in a more appealing form. We have represented in Figure 4.4 (b) the graph G_4 on a three dimensional polyhedron that can be obtained by truncating the six corners of an octahedron. The graphs of Figure 4.4 (a) and 4.4 (b) are topologically the same. The polyhedron of Figure 4.4 (b) has 26 vertices and 36 edges. It is bounded by 6 squares and 8 hexagons and was known to Archimedes and rediscovered by the Russian crystallographer Fedorow. It bears the name of *hexocahedra* or of *tetrakaidekahedron* (see Weyl [122] for more comments on this polyhedron).

As a final remark for this section , let us note that in Theorem 4.8 we assume that the systems p_i do not simultaneously intersect three by three on $\mathbf{R}_{+\infty}$. That is, we assume that there exists no positive real value s_0 for which $p_i(s_0) = p_j(s_0) = p_l(s_0)$ for distinct i, j and l.

The construction of the graph G_4 (and of G_k, $k \geq 5$) can in fact be extended so as to encompass this situation. It suffices therefore to add the edges corresponding to multiple intersections and to label these edges according to the interesctions that they represent.

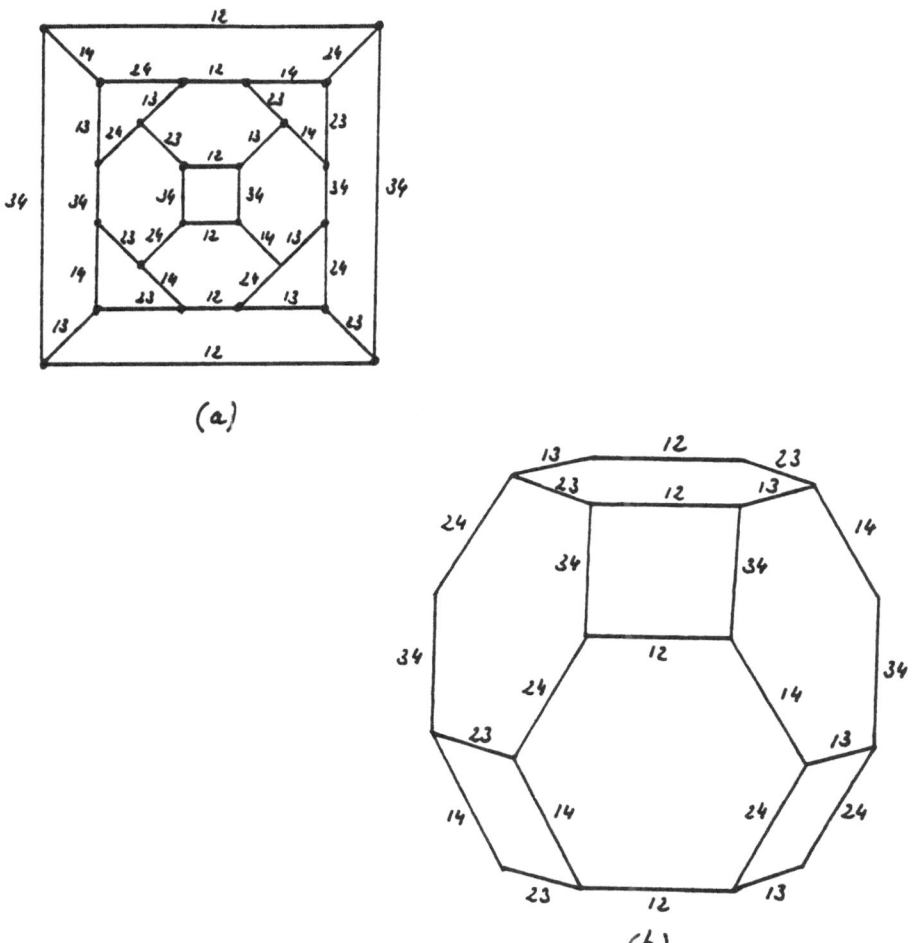

(a)

(b)

Figure 4.4: (a) The graph G_4 for the 4-interlacing property and (b) the same graph represented on a hexocahedra or tetrakaideka-hedron. The systems $p_1(s), p_2(s), p_3(s)$ and $p_4(s)$ satisfy the 4-interlacing condition if and only if the succession of their pairwise intersections, as s increases from 0 to $+\infty$, describes a succession of edges of a path in G_4.

4.4 Intersecting systems

4.4.1 Introduction

The previous section dealt with simultaneous $R_{+\infty}$-stabilization conditions for systems that do not simultaneously intersect. The main theorem of that section (Theorem 4.6) is not valid if $p_i(s_0) = \alpha$ for some $s_0 \in R_{+\infty}$, some $\alpha \in R_\infty$ and all $i = 1, ..., k$, that is if all k systems intersect at $s_0 \in R_{+\infty}$.

In this section we analyse this special case of simultaneous intersection.

From a mathematical point of view it is quite unlikely that arbitrary rational functions $p_i \in R(s)$ all happen to take the same value at the same point of $R_{+\infty}$. From a practical viewpoint, however, this is the generic case. In practice a system is almost always strictly proper and thus it has almost always a zero at infinity. Therefore, in general, $p(\infty) = 0$ and all systems intersect at infinity.

The result of this section is that, when k systems simultaneously intersect somewhere on $R_{+\infty}$, then they are simultaneously $R_{+\infty}$-stabilizable if and only if they are pairwise simultaneously $R_{+\infty}$-stabilizable.

This remarkable situation is really specific to the case of simultaneously intersecting systems: in the previous section we have given an example of three systems that are pairwise simultaneously $R_{+\infty}$-stabilizable but that are not simultaneously $R_{+\infty}$-stabilizable.

We first need the the concept of winding transform and some of its properties.

4.4.2 Winding transform

A real rational function $q \in \mathbf{R}(s)$ maps extended positive real points to values in $\mathbf{R}_\infty$. Such a mapping can be represented on an infinite half plane $\mathbf{R}_{+\infty} \times \mathbf{R}_\infty$ (see Figure 4.5 (a) on page 92). Identifying the two extreme values $+\infty$ and $-\infty$ of the infinite plane we obtain a representation of the mapping q as a curve on a semi-infinite cylinder (see Figure 4.5 (b)). When s increases from 0 to $+\infty$ the curve associated to q 'turns' around the cylinder. It is this notion of 'turn' that we formalise here with the winding transform. But first we introduce the notion of Cauchy index of a rational function.

Definition 4.5 (Gantmacher [48]) *The Cauchy index of a real rational function p on the interval $[a, b]$ ($a, b \in \mathbf{R}_\infty$) is denoted by $I_a^b p$ and is equal to the difference between the number of jumps from $-\infty$ to $+\infty$ and the number of jumps from $+\infty$ to $-\infty$ of $p(s)$ as the argument s changes from a to b.*

One of the methods for computing the Cauchy index of a rational function −without computing its real poles− is based on a theorem of Sturm: see Gantmacher [48] for more details.

The Cauchy index of a rational function on an interval $[a, b]$ is an integer. In light of the representation sketched above it is clearly related to the number of turns that a rational function does on its cylinder representation between a and b. The winding number is a real valued version of the Cauchy index.

Definition 4.6 (Blondel [21]) *Let $a, b \in \mathbf{R}_\infty$ and let $f : [a, b] \to$ $\mathbf{R}_\infty$ be a function that is continuous everywhere on $[a, b]$ except for a finite number of values z_i for which $\lim_{s \to z_i} f(s) = \infty$ (typically we have $f \in \mathbf{R}(s)$). A winding transform of f is a continuous function $F : [a, b] \to \mathbf{R}$ that is such that*

$$\tan(\pi F(s)) = f(s), \quad s \in [a, b].$$

The set of all winding transforms of f is denoted by Wf. The winding number of f on the interval $[a, b]$ is denoted by $W_a^b f$ and is defined by

$$W_a^b f = F(b) - F(a)$$

where F is any winding transform of f. This definition is independent of the choice of the winding transform of f.

Winding transforms always exist. They differ only by an integer and they are, in the sense given below, 'close' to the Cauchy index.

Theorem 4.9 *Assume that $p \in \mathbf{R}(s)$ and let $a, b \in \mathbf{R}_\infty$. Then*

1. *there exists a continuous function P such that $\tan(\pi P(s)) = p(s)$ for all $s \in [a, b]$,*

2. *any two such functions P_1 and P_2 are such that $P_1 = P_2 + k$ for some fixed $k \in \mathbf{Z}$,*

3. *$|I_a^s p + W_a^s p| < 1$ for all $s \in [a, b]$.*

Proof The function defined by $P'(s) = \frac{\arctan p(s)}{\pi}$ is continuous everywhere on $\mathbf{R}_{+\infty}$ except at the poles of p on $\mathbf{R}_{+\infty}$ where it jumps either from $\frac{1}{2}$ to $-\frac{1}{2}$ or from $-\frac{1}{2}$ to $\frac{1}{2}$ depending on whether $p(s)$ jumps from $+\infty$ to $-\infty$ or from $-\infty$ to $+\infty$. In both these cases the continuity is re-established if we add an integer valued function that has a negative (respectively positive) integer jump at the values for which p jumps from $+\infty$ to $-\infty$ (respectively from $-\infty$ to $+\infty$). The Cauchy index $I_a^s p$ of p between a and s has this property. The function defined by

$$P(s) = \frac{\arctan p(s)}{\pi} - I_a^s p$$

is continuous and satisfies

$$\tan(\pi P(s)) = p(s).$$

This proves the first point.

For the second point assume that $\tan(\pi F_1(s)) = \tan(\pi F_2(s)) = p(s)$ for all s in $[a, b]$. Then $\arctan(\tan(\pi F_1(s))) = \arctan(\tan(\pi F_2(s)))$ and thus $\pi F_1(s) = \pi F_2(s) + \pi k(s)$ for some integer valued function $k(s)$. But since F_1 and F_2 are both continuous $k(s)$ must also be continuous, hence point 2 is proved.

For the third point we use the fact that, by a combination of the first two points, any winding transform of $p(s)$ has the form $P(s) = \frac{\arctan p(s)}{\pi} - I_a^s p + k$ for some $k \in \mathbb{Z}$. We then compute

$$|W_a^s p + I_a^s p| = \frac{|\arctan p(s) - \arctan p(a)|}{\pi}$$

and thus, because both $\arctan p(a)$ and $\arctan p(s)$ belong to the interval $(-\frac{\pi}{2}, \frac{\pi}{2}]$, we have

$$|W_a^s p - I_a^s p| < 1.$$

∎

4.4.3 Properties of winding transforms

The next two results show that, associated with any continuous real valued function $c(s) : [a, b] \rightarrow \mathbb{R}$ there is a rational function that

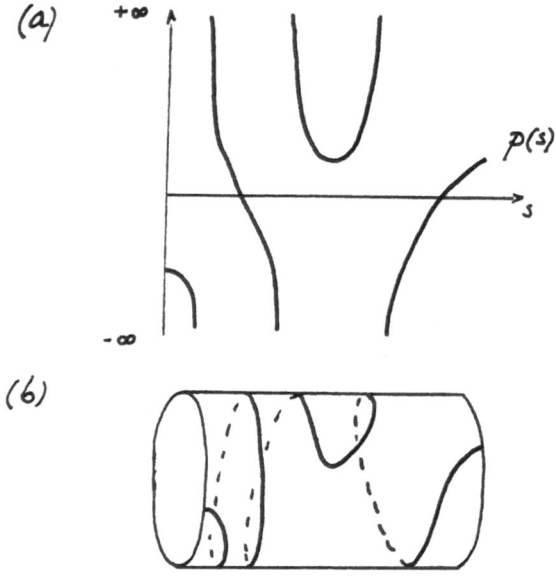

Figure 4.5: (a) The graph of a rational function on the positive real axis and (b) its cylinder representation.

has a winding transform that is arbitrarily close to $c(s)$.

We proceed in two steps. First we prove the result for continuous functions bounded by $\frac{1}{2}$ and then we give a proof of the general case. For convenience we fix $[a, b] = [-1, 1] = I$; the proofs proceed in the same way for other values of a and b.

Theorem 4.10 *Assume that $c(s) : I \to \mathbf{R}$ is continuous and is such that*

$$|c(s)| < \frac{1}{2}, \ s \in I.$$

Then, for any $\epsilon > 0$, there exists a rational function $p \in \mathbf{R}(s)$ that has a winding transform P such that

$$|c(s) - P(s)| < \epsilon, \ s \in I.$$

Proof Define $C(s) \triangleq \tan(\pi c(s))$. Because the absolute value of $c(s)$ is strictly less than $\frac{1}{2}$ on I, the function $C(s)$ is a continuous function on I. By Weierstrass's theorem (polynomials are dense in the set of continuous functions on compact intervals for the supremum norm) there exists a polynomial $t(s) \in \mathbf{R}[s]$ such that

$$\sup_{s \in I} |t(s) - C(s)| < \epsilon.$$

Since $|\arctan a - \arctan b| \leq |a - b|$ we obtain

$$\sup_{s \in I} |\arctan t(s) - \arctan C(s)| \leq \sup_{s \in I} |p(s) - C(s)|.$$

Since

$$\sup_{s \in I} |\arctan t(s) - \pi c(s)| = \sup_{s \in I} |\arctan t(s) - \arctan C(s)|$$

and

$$\sup_{s \in I} |p(s) - C(s)| < \epsilon$$

we finally get

$$\sup_{s \in I} |\arctan t(s) - \arctan C(s)| < \epsilon.$$

The rational function defined by $P(s) \triangleq \frac{\arctan t(s)}{\pi}$ is continuous on I and is such that $\tan(\pi P(s)) = p(s)$. Therefore it is a winding transform of p.

From the last inequality, we have

$$\sup_{s \in I} |\pi P(s) - \pi c(s)| < \epsilon.$$

But then also

$$\sup_{s \in I} |P(s) - c(s)| < \frac{\epsilon}{\pi} < \epsilon$$

as requested. ∎

The same theorem remains valid for continuous functions that have a modulus that is larger than $\frac{1}{2}$ if we add the condition that the number of points on I where $2c(s)$ is integer valued is finite.

Theorem 4.11 *Assume that $c(s)$ is a real valued continuous function on I and suppose that the set T defined by*

$$T \triangleq \{s \in I : 2c(s) \in \mathbf{Z}\}$$

is finite.
Then, for any $\epsilon > 0$, there exists $p \in \mathbf{R}(s)$ that has a winding transform P such that

$$|c(s) - P(s)| < \epsilon, \ s \in I.$$

Proof Define

$$P \triangleq \{s \in I : 2c(s) \text{ is an odd number}\}.$$

By assumption P is finite. We denote

$$P = \{p_0, p_1, ..., p_n\} \text{ with } p_0 < p_1 < ... < p_n.$$

Similarly we define

$$Z = \{s \in I : 2c(s) \text{ is an even number}\}$$

and denote

$$Z = \{z_0, z_1, ..., z_m\} \text{ with } z_0 < z_1 < ... < z_m.$$

We construct the rational function

$$q(s) \triangleq \frac{\prod_{i=0}^{m}(s - z_i)}{\prod_{i=0}^{n}(z - p_i)}.$$

The tangent of this function evaluated on the real positive axis is equal to 0 at z_i and only there and is equal to $\frac{\pi}{2}$ exactly at $s = p_i$. Therefore there exists a winding transform $Q(s)$ of $q(s)$ that is such that

$$|c(s) - Q(s)| < \frac{1}{2}, \ s \in I.$$

But then the function defined by

$$c'(s) \triangleq c(s) - Q(s)$$

is a continuous function that satisfies the conditions of Theorem 4.10, and thus there exists a rational function $r(s)$ together with one of its winding transforms $R(s)$ such that

$$|c'(s) - R(s)| < \epsilon, \ s \in I.$$

That is,

$$|c(s) - (Q(s) + R(s))| < \epsilon, \ s \in I$$

and thus

$$\left| c(s) - \left(\frac{\arctan q(s)}{\pi} + \frac{\arctan r(s)}{\pi} + l_q(s) + l_r(s) \right) \right| < \epsilon, \ s \in I$$

where $l_q(s)$ and $l_r(s)$ are integer valued functions defined on I. Using the fact that

$$\arctan a + \arctan b = \arctan \frac{a + b}{1 - ab} + k\pi \text{ for } k = -1, 0 \text{ or } 1$$

we get

$$\left| c(s) - \frac{\arctan(\frac{q(s) + r(s)}{1 - r(s)q(s)})}{\pi} + k(s) + l_q(s) + l_r(s) \right| < \epsilon,$$

where $k(s)$ is a function that takes the values $-1, 0$ or 1. The expression

$$T(s) \triangleq \frac{\arctan(\frac{q(s)+r(s)}{1-r(s)q(s)})}{\pi} + k(s) + l_q(s) + l_r(s)$$

is a winding transform of $\frac{q(s)+r(s)}{1-r(s)q(s)}$ because it is continuous and such that

$$\tan(\pi T(s)) = \frac{q(s) + r(s)}{1 - r(s)q(s)}.$$

Defining

$$t(s) \triangleq \frac{q(s) + r(s)}{1 - r(s)q(s)},$$

the theorem is then proved. ∎

4.4.4 Intersecting systems

The link between stabilization and winding numbers is as follows.

Theorem 4.12 *Assume that the systems p_1 and p_2 intersect at $s_0 \in \mathbf{R}_{+\infty}$. Then p_1 and p_2 are simultaneously $\mathbf{R}_{+\infty}$-stabilizable if and only if*

$$|W_{s_0}^s p_1 - W_{s_0}^s p_2| < 1 \text{ for all } s \in \mathbf{R}_{+\infty}.$$

Proof We assume without loss of generality that $p_1(s_0) = p_2(s_0) = 0$. (If $p_1(s_0) \neq 0$, define $p_1'(s) \triangleq p_1(s) - p_1(s_0)$ and $p_2'(s) \triangleq p_2(s) - p_2(s_0)$, for which $p_1'(s_0) = p_2'(s_0) = 0$, apply the theorem to $p_1'(s)$ and $p_2'(s)$ and note that p_1 and p_2 are simultaneously $\mathbf{R}_{+\infty}$-stabilizable if and only if p_1', p_2' are.)

For the necessary part, assume by contradiction that $p_1(s_0) = p_2(s_0) = 0$, that $c \in \mathbf{R}(s)$ avoids both p_1 and p_2 on $\mathbf{R}_{+\infty}$ and that

$$|W_{s_0}^{s_1} p_1 - W_{s_0}^{s_1} p_2| \geq 1 \text{ for some } s_1 \in \mathbf{R}_{+\infty}.$$

By continuity of the expression $W_{s_0}^s p_1 - W_{s_0}^s p_2$, there exists some $s_2 \in [s_0, s_1]$ that satisfies

$$|W_{s_0}^{s_2} p_1 - W_{s_0}^{s_2} p_2| = 1.$$

Consider any winding transform $C(s)$ of $c(s)$. The function defined by

$$f(s) \triangleq |(W_{s_0}^s p_1 - C(s)) + (C(s) - W_{s_0}^s p_2)|$$

is continuous and such that $f(s_0) = 0$ and $f(s_2) = 1$. In addition, it is the absolute value of the sum of two continuous functions. Therefore there must exist some integer value $k \in \mathbf{Z}$ and a point $s_3 \in [s_0, s_2]$ for which either

$$W_{s_0}^{s_3} p_1 - C(s_3) = k$$

or

$$W_{s_0}^{s_3} p_2 - C(s_3) = k.$$

We claim that these cases both lead to a contradiction. Indeed, we have

$$\pi W_{s_0}^{s_3} p_i = \pi(k + C(s_3))$$

for either $i = 1$ or $i = 2$. But then, evaluating the tangent on both sides and using the combined facts that $p_i(s_0) = 0$ and $\tan(\pi C(s_3)) = c(s_3)$, we have

$$p_i(s_3) = c(s_3)$$

for either $i = 1$ or $i = 2$.

This contradicts the fact that c avoids both p_1 and p_2 on $\mathbf{R}_{+\infty}$ and thus necessity is proved.

For sufficiency suppose that for some $\epsilon > 0$

$$|W_{s_0}^s p_1 - W_{s_0}^s p_2| < 1 - \epsilon < 1 \text{ for all } s \in \mathbf{R}_{+\infty}.$$

For every $s \in \mathbf{R}_{+\infty}$, define

$$c(s) \triangleq \max\{W_{s_0}^s p_1, W_{s_0}^s p_2\} + \frac{\epsilon}{2}.$$

$c(s)$ is a continuous function such that

$$|c(s) - W_{s_0}^s p_i| > \frac{\epsilon}{2} \text{ for all } s \in \mathbf{R}_{+\infty} \text{ and for } i = 1, 2$$

and

$$|c(s) - W_{s_0}^s p_i| < 1 - \frac{\epsilon}{2} \text{ for all } s \in \mathbf{R}_{+\infty} \text{ and for } i = 1, 2.$$

By Theorem 4.11 there exists a $q \in \mathbf{R}(s)$ such that

$$|c(s) - W_{s_0}^s q| < \frac{\epsilon}{4} \text{ for all } s \in \mathbf{R}_{+\infty}.$$

For this q we then have,

$$\frac{\epsilon}{4} < |W_{s_0}^s q - W_{s_0}^s p_i| < 1 - \frac{\epsilon}{4} \text{ for all } s \in \mathbf{R}_{+\infty} \text{ and for } i = 1, 2.$$

This shows that q avoids p_1 and p_2 on $\mathbf{R}_{+\infty}$, hence the result. ∎

The next final result is proved by using the winding transform.

Theorem 4.13 *Assume that k systems $p_i \in \mathbf{R}(s)$ $(i = 1, ..., k)$ simultaneously intersect somewhere on $\mathbf{R}_{+\infty}$. Then p_i are simultaneously $\mathbf{R}_{+\infty}$-stabilizable if and only if they are pairwise simultaneously $\mathbf{R}_{+\infty}$-stabilizable.*

Proof Necessity is immediate. We show sufficiency only.
Assume that the systems simultaneously intersect at $s_0 \in \mathbf{R}_{+\infty}$ and assume, without loss of generality, that they all take the value

zero there. (If not, define $p'_i = p_i - p_i(s_0)$, apply the theorem to p'_i for which $p'_i(s_0) = 0$ and note that p_i are simultaneously $\mathbf{R}_{+\infty}$-stabilizable if and only if p'_i are.)

Since the systems are pairwise simultaneously stabilizable it follows by the previous theorem that, for some $\epsilon > 0$

$$|W^s_{s_0}p_i - W^s_{s_0}p_j| < 1 - \epsilon < 1, \quad \text{for all } s \in \mathbf{R}_{+\infty}, \ (i,j = 1, ..., k).$$

We construct a continuous function in the following way. For every $s \in \mathbf{R}_{+\infty}$, we define

$$c(s) = \max_{i=1,...,k}\{W^s_{s_0}p_i(s)\} + \frac{\epsilon}{2}.$$

c is continuous and such that

$$\frac{\epsilon}{2} < |c(s) - W^s_{s_0}p_i(s)| < 1 - \frac{\epsilon}{2} \quad \text{for all } s \in \mathbf{R}_{+\infty}, \ (i,j = 1, ..., k).$$

By Theorem 4.11 there exists a $q \in R(s)$ that is such that

$$|c(s) - W^s_{s_0}q| < \frac{\epsilon}{4} \quad \text{for all } s \in \mathbf{R}_{+\infty}.$$

For this rational function q we have

$$\frac{\epsilon}{4} < |W^s_{s_0}q - W^s_{s_0}p_i| < 1 - \frac{\epsilon}{4}, \quad \text{for all } s \in \mathbf{R}_{+\infty}, \ (i = 1, ..., k).$$

This shows that $q \in R(s)$ avoids $p_i(s)$ $(i = 1, ..., k)$ on $\mathbf{R}_{+\infty}$ and the theorem is proved. ∎

Corollaries of this theorem include:

Corollary 4.1 *Assume that k systems $p_i \in R(s)$ $(i = 1, ..., k)$ simultaneously intersect at $s_0 \in \mathbf{R}_{+\infty}$. Then they are simultaneously $\mathbf{R}_{+\infty}$-stabilizable if and only if*

$$|W^s_{s_0}p_i - W^s_{s_0}p_j| < 1 \text{ for all } s \in \mathbf{R}_{+\infty} \text{ and } i, j \in \{1, 2, ..., k\}.$$

Corollary 4.2 *Assume that $p_i \in R(s)$ $(i = 1, ..., k)$ have a common pole or zero on $R_{+\infty}$. Then they are simultaneously $R_{+\infty}$-stabilizable if and only if they are pairwise simultaneously $R_{+\infty}$-stabilizable.*

Corollary 4.3 *Assume that $p_i \in R(s)$ $(i = 1, ..., k)$ are strictly proper. Then they are simultaneously $R_{+\infty}$-stabilizable if and only if they are pairwise simultaneously $R_{+\infty}$-stabilizable.*

4.5 Summary and bibliography

In this chapter we have identified $R_{+\infty}$-stabilization conditions as important necessary conditions for stabilization and we have conducted a long analysis of $R_{+\infty}$-stabilization.

For the case of two systems there exist necessary and sufficient conditions in the form of interlacing conditions between poles and zeros on $R_{+\infty}$. For more than two systems, two cases must be taken into account.

The simplest case is the one for which the systems simultaneously intersect somewhere on $R_{+\infty}$. The systems are then simultaneously stabilizable if and only if they are pairwise simultaneously $R_{+\infty}$-stabilizable. The proof of this result made an astute use of the concept of winding number and of winding transform. Since checkable conditions exist for pairwise $R_{+\infty}$-stabilization, one conclude that a complete solution exists for simultaneous $R_{+\infty}$-stabilization of k systems that intersect somewhere on $R_{+\infty}$. A special case of this is the case of strictly proper systems.

When the systems do not simultaneously intersect on $\mathbf{R}_{+\infty}$ the picture is different and we have provided a family of interlacing conditions for these situations: the even interlacing condition for bistable $\mathbf{R}_{+\infty}$-stabilization and the k-interlacing conditions for simultaneous $\mathbf{R}_{+\infty}$-stabilization of k systems.

The parity interlacing property was first given by Youla *et al.* [123]. The even interlacing condition is from Wei [118] who also conjectured the sufficiency of the condition for bistable stabilization. This conjecture is answered negatively in the next chapter. The 3-interlacing condition is given under a different but equivalent form in Wei [119]. It also appears under a more embryonic form in Ghosh [52]. The k-interlacing condition appears for the first time in Blondel [16]. Winding numbers (see Blondel [21]) are an adaptation of the well established Cauchy index [48] and of a geometrical concept introduced in Brockett [23]. It also has a Hilbert transform flavour: see Anderson, Dasgupta, Khargonekar, Krause and Mansour [6] and references therein for more details.

Chapter 5

Sufficient conditions: special cases

Je cherche à comprendre.

J. Monod.

5.1   Introduction

In the previous chapter we have concentrated our attention on the positive real axis and this has allowed us to answer completely the question of simultaneous $\mathbf{R}_{+\infty}$-stabilization. The interlacement conditions obtained are necessary conditions for the true question of simultaneous stabilization. In this chapter the sufficiency of these conditions is analysed and two situations for which systems are simultaneously stabilizable are proposed. Throughout this chapter we mean $C_{+\infty}$-stability when making no explicit reference to any instability region.

In Section 2 it is shown that the conditions obtained for strong $R_{+\infty}$-stabilization and for simultaneous $R_{+\infty}$-stabilization of two systems are also sufficient for strong and simultaneous stabilization: if there exists a controller that is such that the closed loop transfer functions associated to p_1 and p_2 have no real unstable poles, then there also exists a controller that is such that these closed loop transfer functions have no unstable poles at all.

In Section 3 we show that this striking property of equivalence between $R_{+\infty}$- and $C_{+\infty}$-stabilization of two systems is no longer true for three or more systems or for bistable stabilization.

The first two examples of Section 3 present of four and three systems that are simultaneously $R_{+\infty}$-stabilizable but not simultaneously stabilizable. The last example is on bistable stabilization. We give an example of a system that is $R_{+\infty}$-stabilizable by a $R_{+\infty}$-bistable controller but that is not stabilizable by a bistable controller. This last statement also shows that a system that is stabilizable by a stable controller and stabilizable by an inverstable controller is not always stabilizable by a stable *and* inverstable (bistable) controller. The proofs of all these results use advanced properties of the range of analytic functions that are reviewed in the appendices B and C.

Section 4 is on sufficient conditions for simultaneous stabilization. In particular it is shown that simultaneous stabilzation of k systems can always be achieved if one of the systems avoids all the others in $C_{+\infty}$.

The robust control techniques in system theory can be employed to derive tractable sufficient conditions for simultaneous stabilization. The H_∞ approach to robust control is particularly effective for this purpose. In Section 5 we briefly overview this methodology and put it into service for simultaneous stabilization.

5.2 Two systems and strong stabilization

The objective of this section is to show that a system that is strongly $R_{+\infty}$-stabilizable is also strongly stabilizable.

From the results contained in Chapter 4 we know that a system is strongly $R_{+\infty}$-stabilizable if and only if it satisfies the parity interlacing condition. What remains to be shown is thus that the parity interlacing condition is sufficient for strong stabilization.

Strong stabilization conditions rely on interpolating conditions by a bistable rational function.

Theorem 5.1 *Assume that $p \in R(s)$ and let $p = \frac{n_p}{d_p}$ be a fractional factorization of p in S. Then p is strongly stabilizable if and only if there exists $u \in U$ such that,*

$$u(s) = d_p(s) \text{ for all } s \in C_{+\infty} \text{ for which } n_p(s) = 0.$$

Proof By Theorem 2.6, p is strongly stabilizable if and only if there exists an element $c \in S$ such that

$$n_p c + d_p = u \in U.$$

This equation has a solution for some $c \in S$ if and only if

$$n_p c = u - d_p$$

for some $u \in U$ and $c \in S$. This is solvable if and only if

$$u - d_p \text{ is divisible by } n_p \text{ for some } u \in U.$$

By Theorem 2.2 this divisibility condition can be satisfied if and only if $u(s) - d_p(s)$ is equal to zero in $\mathbf{C}_{+\infty}$ whenever $n_p(s)$ is, i.e. if and only if

$$u(s) = d_p(s) \text{ when } n_p(s) = 0$$

for $s \in \mathbf{C}_{+\infty}$. ∎

We illustrate the link between strong stabilization and interpolation with a short example.

A fractional factorization of the system $p = \frac{s}{(s-1)(s-2)}$ in S is given by $n_p = \frac{s}{(s+1)^2}$ and $d_p = \frac{(s-1)(s-2)}{(s+1)^2}$. The zeros of n_p in $\mathbf{C}_{+\infty}$ are at $s = 0$ and at $s = \infty$, and thus, p is strongly stabilizable if and only if there exists $u \in U$ such that $u(0) = d_p(0) = 2$ and $u(\infty) = d_p(\infty) = 1$. These two conditions are simultaneously satisfied by $u = \frac{s+2}{s+1}$ and thus the system $p = \frac{s}{(s-1)(s-2)}$ is strongly stabilizable.

Bistable rational functions are continuous, real valued and are never equal to zero on the real positive axis. Therefore, they must always have the same sign on the real positive axis: either positive or negative. It is not possible to satisfy, say, $u(0) = 2$ and $u(\infty) = -1$ for $u(s) \in U$. This 'same sign' condition is, remarkably, also sufficient for the interpolation problem above: a bistable rational function can interpolate any given set of non-zero values at any given set of points of the right half plane as long as the values associated to the real points are real and all have the same sign.

The first historical proof of this result was constructive but is tedious and relatively long (see Youla [123]). We give hereafter a more advanced but shorter proof.

Theorem 5.2 (Vidyasagar [108]) *Let (s_i, σ_i) $(i = 1, ..., n)$ be n pairs in $\mathbf{C}_{+\infty} \times \mathbf{C}$ such that*

$$s_i = \bar{s}_j \ (i, j = 1, ..., n) \Rightarrow \sigma_i = \bar{\sigma}_j.$$

Assume, without loss of generality, that s_i $(i = 1, ..., m)$ are real and that s_i $(i = m + 1, ..., n)$ are not. Then there exists an interpolating bistable rational function $u \in U$ such that

$$u(s_i) = \sigma_i, \ (i = 1, ..., n),$$

if and only if

- *σ_i are non zero for $i = 1, ..., n$,*
- *σ_i are real and all have the same sign when $i = 1, ..., m$.*

Note that the second constraint is on the m first values only.

Proof Necessity is obvious: $u \in U$ is real valued, continuous and is never equal to zero on $\mathbf{R}_{+\infty}$. Hence it must have the same sign everywhere on $\mathbf{R}_{+\infty}$ and so for the interpolating values associated to points on $\mathbf{R}_{+\infty}$.

Sufficiency is the difficult part. Assume without loss of generality that the signs of σ_i $(i = 1, ..., m)$ are all positive. (If not, define

$\sigma'_i = -\sigma_i$, find a bistable rational function u' that interpolates the positive values σ'_i at the points s_i $(i = 1, ..., n)$ and define $u = -u'(s)$. u satisfies the interpolating constraints.)

First, using the bilinear transformation from $C_{+\infty}$ to $\overline{D}$: the interpolation conditions are transferred in $\overline{D}$. The interpolating points are mapped from $C_{+\infty}$ to $\overline{D}$ according to

$$s_i \rightarrow z_i = \frac{s_i - 1}{s_i + 1} \ (i = 1, ..., n)$$

whereas the interpolating values are concerved as they are. Due to the properties of the bilinear mapping ($\sigma(s)$ maps the extended real line $R_{+\infty}$ on the interval $I = [-1, 1]$), the points z_i $(i = 1, ..., m)$ are real whereas z_i $(i = m + 1, ..., n)$ are not.

Second, define

$$\delta_i = \ln \sigma_i \ (i = 1, ..., n).$$

Note that δ_i are real numbers for $i = 1, ..., m$ because σ_i are all positive for $i = 1, ..., m$ and that logarithms of positive numbers are real. It is here that the proof fails if σ_i $(i = 1, ..., m)$ do not all have the same sign.

Third, find a real interpolating polynomial $p(z) \in R[z]$ such that

$$p(z_i) = \delta_i \ (i = 1, ..., n).$$

This is a classical interpolation problem; it is easy to perform and can always be achieved by using, for example, the Lagrange interpolation method. Define therefore

$$p(z) = \sum_{j=1}^{n} \left(\frac{\prod_{i=1}^{n}(z - z_i)}{\prod_{i=1,i\neq j}^{n}(z_j - z_i)} \frac{\delta_j}{z - z_j} \right)$$

. The polynomial may is real.

Fourth, define

$$f(z) = e^{p(z)}.$$

This expression is valid because $p(z)$ is a member of the Banach algebra $A(\overline{D})$ of analytic functions in D that are continuous on $\overline{D}$ and exponentials in Banach algebras are well defined (see Appendix B). Thus $f(z)$ is a member of $A(\overline{D})$.

Note that, because $e^{p(z)}e^{-p(z)} = 1$, $f(z)$ is invertible in $A(\overline{D})$ and that, therefore, it never takes the value zero in $\overline{D}$. Note also that

$$f(z_i) = e^{p(z_i)} = e^{\delta_i} = e^{\ln \sigma_i} = \sigma_i \ (i = 1, ..., n).$$

The analytic function $f(z)$ is never equal to zero in $\overline{D}$ and it satisfies the required interpolation constraints. It remains to show that the same objectives can be achieved with a *bistable rational* function. This is what is done in the next two steps.

Fifth, define $q(z)$ to be equal to any real polynomial such that

$$q(z_i) = \sigma_i \ (i = 1, ..., n).$$

Again, this is done by using, for example, the Lagrange interpolation method.

The function $g(z)$ defined by

$$g(z) = \frac{f(z) - q(z)}{\prod_{i=1}^{m}(z - z_i) \prod_{i=m+1}^{n}(z - z_i)(z - \overline{z}_i)}$$

is then a valid member of $A(\overline{D})$ because the poles z_i all cancel with the zeros of $f(z) - q(z)$ since $f(z_i) = q(z_i)$ and $f(\overline{z}_i) = q(\overline{z}_i)$ for $i = 1, ..., n$.

By density of $S(\overline{D})$ in $A(\overline{D})$ (see Appendix B) find $r(z) \in S(\overline{D})$ sufficiently close to $g(z)$ so that the rational function defined by

$$v(z) = q(z) + r(z) \prod_{i=1}^{m}(z - z_i) \prod_{i=m+1}^{n}(z - z_i)(z - \overline{z}_i)$$

is

- never equal to zero in $\overline{D}$,

- rational and with no poles in $\overline{D}$,

- such that $v(z_i) = q(z_i) = \sigma_i$ for $i = 1, ..., n$.

Sixth, use again the bilinear transformation and define

$$u(s) = v(\frac{s-1}{s+1}) \in \mathbf{R}(s).$$

The bistable rational function u satisfies all the requirements of the theorem and the proof is complete. ∎

To keep the notations at a reasonable level we have excluded the case for which the successive derivatives of u must also interpolate specified values at specified points of the right half plane. The theorem remains correct with these supplementary interpolation constraints on the derivatives and without any additional hypothesis, save for the trivial one that the derivative(s) at a point on the real axis must be real. The proof follows exactly in the same way in this case.

The proof is not constructive since we compute the exponential of a polynomial at the fourth step and this computation needs the evaluation of an infinite sum (the exponential of $p(z) \in \mathbf{R}[z]$ is defined by the convergent series $e^{p(z)} = \lim_{n \to \infty} \sum_{i=1}^{n} \frac{p(z)^i}{i!}$).

Nothing in the proof gives the slighest indication about the degree of the interpolating bistable rational function. This issue is important since this degree is directly connected to that of the strong stabilizing controller of Theorem 5.1. Some comments about this topic are given in the bibliography of this chapter.

A direct application of Theorem 5.2 is the next result.

Theorem 5.3 *The system $p \in \mathbf{R}(s)$ is strongly stabilizable if and only if it satisfies the parity interlacing condition.*

Proof The proof follows from the last two results.

Let $\frac{n_p}{d_p}$ be a fractional factorization of p in S. By Theorem 5.1, p is strongly stabilizable if and only if there exists a bistable rational function $u \in U$ such that $u(s) = d_p(s)$ when $n_p(s) = 0$ for $s \in \mathbf{C}_{+\infty}$. By Theorem 5.2 this interpolation constraint can be satisfied if and only if $d_p(s)$ always has the same sign at the *real* points $s \in \mathbf{C}_{+\infty}$ for which $n_p(s) = 0$. This condition is again satisfied if and only if d_p has an even number of zeros on $\mathbf{R}_{+\infty}$ between each pair of zeros of n_p on $\mathbf{R}_{+\infty}$. Zeros of n_p and d_p on $\mathbf{R}_{+\infty}$ are zeros and poles of p on $\mathbf{R}_{+\infty}$; hence the result. ∎

The parity interlacing property is necessary and sufficient for strong $\mathbf{R}_{+\infty}$-stabilization and for strong stabilization. Therefore:

Corollary 5.1 *A system is strongly $\mathbf{R}_{+\infty}$-stabilizable if and only if it is strongly stabilizable.*

A similar formulation holds for simultaneous stabilization of two systems.

Theorem 5.4 *Two systems are simultaneously $\mathbf{R}_{+\infty}$-stabilizable if and only if they are simultaneously stabilizable.*

Proof Sufficiency is obvious.

For necessity, assume that $p_1, p_2 \in \mathbf{R}(s)$ and let $\frac{n_1}{d_1}$ and $\frac{n_2}{d_2}$ be fractional factorizations of p_1 and p_2 in S. Consider $x, y \in S$ such that

$$n_1 x + d_1 y = 1.$$

Since p_1 and p_2 are simultaneously $\mathbf{R}_{+\infty}$-stabilizable, Theorem 4.1 implies that

$$p \triangleq \frac{n_1 x + d_1 y}{n_1 d_2 - n_2 d_1}$$

is strongly $\mathbf{R}_{+\infty}$-stabilizable. By using Corollary 5.1 we infer that p is strongly stabilizable. But then, using Theorem 4.1 again, it follows that p_1 and p_2 are simultaneously stabilizable as requested. ∎

Before proceeding to the next section we make one more comment. The parity interlacing condition provides a nice, effective and tractable criterion to decide whether or not a system is stabilizable by a stable controller. Because of the exponential in the proof given for Theorem 5.2, the procedure is not constructive and it is thus helpless for the construction of a stable stabilizing controller when such a controller exists. There exist alternative constructive procedures for computing stable stabilizing controllers. One of their important common drawbacks, however, is that they may, and they do, lead to high degree controllers. Their practical interest is therefore quite limited. The problem of finding, when one exists, a stable stabilizing controller of lowest possible degree is an important open question. Even then, a 'minimal degree' procedure may be inacceptable from a practical point of view because there exists intrinsic limitations on the lowest possible order of a stable stabilizing controller. When a stable stabilizing controller exists, it may well be unacceptable, from a practical point of view, to construct such a controller because its degree is not bounded by the degree of the system. There exist examples of systems that have a degree equal to two, that are strongly stabilizable, but for which any stable stabilizing controller is of degree, say, at least 1995 (see e.g. Ghosh [50]).

This short discussion is for strong stabilizing controllers but it carries over to the case of simultaneous stabilization of two systems.

5.3 Sufficiency of interlacement conditions

5.3.1 Introduction

The message of this section is that, unlike the case of two systems, three or more systems that are simultaneously $R_{+\infty}$-stabilizable are not automatically simultaneously stabilizable.

This property is shown by means of examples. By a careful choice of special systems we show that if the above property was true it would then contradict theorems on the range of analytic functions.

Classical results from analytic function theory which we use in this section are contained in Appendix B (maximum modulus theorem, Rouché's theorem, Hurwitz's theorem, generalized form of Montel's criterion for normal families). The other theorems that we need are detailed in a special appendix that discloses remarkable properties on the range of analytic functions. These results were collected from various sources and we have endeavoured to present them in a clear and logical way in appendix C. Because of their crucial importance for the proofs of the theorems of this section, we strongly recommend that the reader browse through Appendix C before reading this section.

We start with an example of four systems that are simultaneously **R**$_{+\infty}$-stabilizable but not simultaneously stabilizable. This example uses Picard-Schottky's Theorem. The second example is of the same kind but for three systems and its proof needs the $\frac{1}{16}$-Theorem. The last example is of a system that is bistably **R**$_{+\infty}$-stabilizable but not bistably stabilizable. There we need Montel's Theorem.

These examples need quite different results in analytic function theory and we feel that, for this reason, they all deserve an exposition in this monograph even though they provide counterexamples to the same conjecture.

For consistency with the rest of the monograph all examples are given in a **C**$_{+\infty}$-stability set-up rather than in their more natural $\overline{D}$-stability context.

5.3.2 Four systems

The next proof uses Picard-Schottky's theorem.

Theorem 5.5 (Picard-Schottky) *A function that is analytic in D and that never takes the value 0 nor the value 1 on D is bounded in D by*

$$|f(z)| \leq (e^\pi \max\{|f(0)|, 1\})^{\frac{1+|z|}{1-|z|}}, \; z \in D.$$

■

Theorem 5.6 *The four systems* $p_1(s) = 0, p_2(s) = \frac{s-1}{s+1}, p_3(s) = 2\frac{s-1}{s+1}$ *and* $p_{4,\delta}(s) = \frac{s-1}{(1-\delta)s+(1+\delta)}$ *are simultaneously* **R**$_{+\infty}$*-stabilizable for every* $\delta > 0$ *but they are not simultaneously stabilizable when* $\delta > \overline{\delta} = 8e^{3\pi} = 99105.$

Proof When $s = 1 \in \mathbf{R}_{+\infty}$ all four systems take the value 0, i.e. they simultaneously intersect at the unstable point $s = 1$. Additionally, it is easy to check that the systems are pairwise simultaneously stabilizable and thus, applying Theorem 4.13, the systems p_1, p_2, p_3 and p_4 are simultaneously $\mathbf{R}_{+\infty}$-stabilizable. The first part of the proof, the easy one, is thus proved: the four systems are simultaneously $\mathbf{R}_{+\infty}$-stabilizable for every $\delta > 0$.

It remains to show that, when $\delta > \overline{\delta}$, the systems are not simultaneously stabilizable. In order to do this we first translate the problem into a $\overline{D}$-stability framework.

Using the bilinear transformation $z = \frac{s-1}{s+1}$ and its inverse $s = \frac{1+z}{1-z}$, the four systems are seen to be simultaneously stabilizable if and only if the four systems defined by $p_1'(z) = 0, p_2'(z) = z, p_3'(z) = 2z$ and $p_{4,\delta_0}'(z) = \frac{\frac{1+z}{1-z}-1}{(1-\delta)\frac{1+z}{1-z}+(1+\delta)} = \frac{z}{1-\delta z}$ are simultaneously $\overline{D}$-stabilizable. Suppose, by contradiction, that, for some $\delta_0 > \overline{\delta}$, we have a $\overline{D}$-stabilizing controller $c(z)$ of fractional decomposition $c(z) = \frac{n_c(z)}{d_c(z)}$, $n_c(z), d_c(z) \in \mathbf{R}[z]$ for the systems $p_1(z), p_2(z), p_3(z)$ and $p_{4,\delta_0}(z)$.

Since $c(z)$ stabilizes $p_1(z) = 0$, $c(z)$ must be $\overline{D}$-stable and so $d_c(z)$ has no zeros in $\overline{D}$. The fact that $c(z)$ also stabilizes $p_2(z), p_3(z)$ and $p_{4,\delta_0}(z)$ leads to the following three inclusions

$$zc(z) + 1 \in U(\overline{D}),$$
$$2zc(z) + 1 \in U(\overline{D}),$$
$$zc(z) + 1 - \delta_0 z \in U(\overline{D}).$$

Defining $f(z) \triangleq 2(zc(z)+1) \in S(\overline{D})$, the inclusions can be rewritten as

$$\frac{1}{2}f(z) \in U(\overline{D}),$$
$$f(z) - 1 \in U(\overline{D}),$$

$$\frac{1}{2}f(z) - \delta_0 z \ \in U(\overline{D}).$$

We now apply Picard-Schottky's theorem. Note first that $f(z)$ is analytic in the open unit disc and that $f(0) = 2$. In addition we have from the first two inclusions

$$f(z) \neq 0, \ z \in D$$

and

$$f(z) \neq 1, \ z \in D.$$

The third one reads

$$f(z) - 2\delta_0 z \neq 0, \ z \in D.$$

The first two inclusions express that 0 and 1 are lacunary values of $f(z)$ in D, i.e. $f(z) \neq 0$ and $f(z) \neq 1$ when $z \in D$. In the sequel we show that this, together with Picard-Schottky's theorem, contradicts the third inclusion. Indeed, due to the fact that 0 and 1 are lacunary values of $f(z)$ in D, we know by Picard-Schottky's theorem that

$$|f(z)| \leq (e^\pi \max\{|f(0)|, 1\})^{\frac{1+|z|}{1-|z|}}, \ z \in D.$$

That is,

$$|f(z)| \leq (2e^\pi)^{\frac{1+|z|}{1-|z|}}, \ z \in D.$$

In particular, when considering $|z| = \frac{1}{2}$

$$|f(z)| \leq (2e^\pi)^3 = 8e^{3\pi} = \overline{\delta}, \ |z| = \frac{1}{2}.$$

On the other hand, the function $g(z)$ defined by $g(z) \triangleq 2\delta_0 z$, when evaluated on $|z| = \frac{1}{2}$, is such that

$$|g(z)| = \delta_0.$$

Hence, because $\delta_0 > \overline{\delta}$, we have on $|z| = \frac{1}{2}$

$$|f(z)| < |g(z)|.$$

By Rouché's theorem this leads to the conclusion that the function

$$f(z) - g(z) = f(z) - 2\delta_0 z$$

has the same number of zeros in the ball $B(0, \frac{1}{2})$ as $g(z)$. But $g(z)$ has precisely one zero there, namely at $z = 0$. And thus $f(z) - 2\delta_0 z$ must have a zero in the disc of center $z = 0$ and of radius $\frac{1}{2}$. But this contradicts our third inclusion

$$f(z) - 2\delta_0 z \neq 0, \; z \in D.$$

A contradiction is obtained and the theorem is proved. ∎

5.3.3   Three systems

The next proof uses the $\frac{1}{16}$-Theorem.

Theorem 5.7 ($\frac{1}{16}$-Theorem) *A function that is analytic in D, that is equal to zero at zero but nowhere else in D and whose first derivative at the origin is equal to 1, has a range that contains a ball centered at $z = 0$ and of radius $\frac{1}{16}$ but not always a larger ball.* ∎

Theorem 5.8 *The systems $p_1'(s) = 0, p_2'(s) = \frac{s-1}{s+1}$ and $p_{3,\delta}'(s) = \frac{(s-1)^2}{(1-\delta)s^2 - 2\delta s - (1+\delta)}$ are simultaneously $\mathbf{R}_{+\infty}$-stabilizable for every $\delta > 0$ but they are not simultaneously stabilizable when $\delta < \frac{1}{16}$.*

Proof Using the bilinear transormation we obtain the equivalent formulation in a $\overline{D}$-stability set-up: the systems $p_1(z) = 0, p_2(z) = z$ and $p_{3,\delta}(z) = \frac{z^2}{z-\delta}$ are simultaneously I-stabilizable for any $\delta > 0$ but not simultaneously $\overline{D}$-stabilizable for $\delta < \frac{1}{16}$.

Observe first that $p_1(0) = p_2(0) = p_{3,\delta}(0) = 0$ for any strictly positive δ. Thus the systems simultaneously intersect at $z = 0$. It is also easy to check that the systems are pairwise simultaneously I-stabilizable for any $\delta > 0$ and thus, by Theorem 4.13, the systems are simultaneously I-stabilizable.

This easy part is thus proved and it remains only to show that, when $0 < \delta < \frac{1}{16}$, the systems are not simultaneously $\overline{D}$-stabilizable. We show this by using a contradiction argument.

Assume by contradiction that, for some $0 < \delta_0 < \frac{1}{16}$, there exists a stabilizing controller $c(z) \in \mathbf{R}(z)$ for the three systems $p_1(z), p_2(z)$ and $p_{3,\delta_0}(z)$. Then, since $c(z)$ stabilizes $p_1(z) = 0$, the controller must be $\overline{D}$-stable. But $c(z)$ also stabilizes $p_2(z) = z$ and $p_{3,\delta_0}(z) = \frac{z^2}{z-\delta_0}$ and hence it must also satisfy the two equations,

$$
\begin{aligned}
zc(z) + 1 &= u_1(z), \\
z^2 c(z) + z - \delta_0 &= u_2(z)
\end{aligned}
$$

for some $u_1(z), u_2(z) \in U(\overline{D})$. Let us define

$$f(z) \triangleq z^2 c(z) + z = z(zc(z) + 1) \in S(\overline{D}).$$

Then $f(0) = 0$ and $f'(0) = 1$ but also, due to the above two equations

$$
\begin{aligned}
f(z) &= zu_1(z), \\
f(z) - \delta_0 &= u_2(z).
\end{aligned}
$$

In more geometrical terms the first of these equations means that $f(z)$ is equal to zero at zero and only there while the second equation means that δ_0 is not in the range of $f(z)$. Combined with the fact that $f(0) = 0$ and that $f'(0) = 1$ this contradicts the $\frac{1}{16}$-Theorem.

Indeed, $f(z)$ satisfies the hypothesis of the $\frac{1}{16}$-Theorem and so has a ball of radius $\frac{1}{16}$ in its range. But its range does not contain $\delta_0 < \frac{1}{16}$. This is clearly impossible, a contradiction is obtained and our theorem is proved. ∎

Another comment on our results is the following. We proved in Theorem 5.8 that, when $0 < \delta < \frac{1}{16}$, the three systems $p_1(z) = 0, p_2(z) = z$ and $p_{3,\delta}(z) = \frac{z^2}{z-\delta}$ are not simultaneously $\overline{D}$-stabilizable. This result can actually be made more precise: the systems $p_1(z) = 0, p_2(z) = z$ and $p_{3,\delta}(z) = \frac{z^2}{z-\delta}$ are simultaneously $\overline{D}$-stabilizable *if and only if* $\delta = 0$ or $|\delta| > \frac{1}{16}$. See Blondel [21] for a proof of this.

5.3.4   Bistable stabilization

Finally we give an example of a system that has the even interlacing property (i.e. that is bistably $\mathbf{R}_{+\infty}$-stabilizable) but that is not bistably stabilizable. This is the last and most complicated example. The long proof of this result uses the notion of normal families of analytic functions and the generalized form of Montel's normal family criteria. As we noted already in the introduction of this chapter, this system is then stabilizable by a stable controller, stabilizable by a inverstable controller but not stabilizable by a stable *and* inverstable controller.

Theorem 5.9 *There exists a real number δ with $|\delta| < 1$ that is such that the system*

$$p_\delta(s) = \frac{s^2 - 1}{s^2 - 2\delta s + 1}$$

is not bistably stabilizable although it satisfies the even interlacing condition.

Proof For the even interlacing condition, note that the zeros of $p_\delta(s)$ are at $s = \pm 1$ whereas the poles are at $s = \delta \pm \sqrt{\delta^2 - 1}$. When $|\delta| < 1$, the poles have a non-zero imaginary part and thus the system $p_\delta(s)$ has no poles on the real positive axis. It thus trivially satisfies the even interlacing condition. By Theorem 4.7 this means that $p_\delta(s)$ is bistably $\mathbf{R}_{+\infty}$-stabilizable for all $|\delta| < 1$. This first part is thus proved.

Now comes the difficult portion of the proof. We have to show that there exists some value δ with $|\delta| < 1$ for which the system $p_\delta(s)$ is not bistably stabilizable.

Note first that, using the bilinear transformation we get

$$p_\delta\left(\frac{z+1}{z-1}\right) = \frac{(\frac{z+1}{z-1})^2 - 1}{(\frac{z+1}{z-1})^2 - 2\delta(\frac{z+1}{z-1}) + 1} = \frac{z}{z^2 + \frac{1-\delta}{1+\delta}}.$$

Define $\epsilon = \frac{1-\delta}{1+\delta}$ and note that when $|\delta| < 1$ then $\epsilon > 0$. Due to the equivalence between $\overline{D}$ and $\mathbf{C}_{+\infty}$-stabilizability, we thus have to prove that the system $p(z) = \frac{z}{z^2+\epsilon}$ is not bistably $\overline{D}$-stabilizable , for some strictly positive value of ϵ. We proceed by contradiction. Assume that for every $\epsilon > 0$ there exists a bistable stabilizing controller $u_\epsilon(z)$. In particular, for every positive integer n there exists a bistable stabilizing controller $u_{\frac{1}{n}}(z)$ for $p_{\frac{1}{n}}(z) = \frac{z}{z^2+\frac{1}{n}}$. Then $u_{\frac{1}{n}}(z)$ satisfies

$$u_{\frac{1}{n}}(z)z + z^2 + \frac{1}{n} \in U(\overline{D}).$$

Then also, multiplying this expression by n

$$nu_{\frac{1}{n}}(z)z + nz^2 + 1 = z(nu_{\frac{1}{n}}(z) + nz) + 1 \in U(\overline{D}).$$

Defining $f_n(z) \triangleq nu_{\frac{1}{n}}(z) + nz$, we have from the last identity

$$zf_n(z) + 1 \in U(\overline{D}),$$

and from the definition of $f_n(z)$

$$f_n(z) - nz = nu_{\frac{1}{n}}(z) \in U(\overline{D}).$$

In the next part we show that the existence, for every positive integer n, of a simultaneous solution $f_n(z)$ to these two equations is impossible.

We define $g_n(z) \triangleq \frac{z(f_n(z) - nz)}{z f_n(z) + 1} \in S(\overline{D})$. The functions $g_n(z)$ are analytic in D, they have no zeros in $D \setminus \{0\}$ and they take the value 1 only twice in D, namely at $z = \frac{i}{\sqrt{n}}$ and $z = -\frac{i}{\sqrt{n}}$. This last property is due to the fact that

$$\frac{z(f_n(z) - nz)}{z f_n(z) + 1} = 1 \Leftrightarrow nz^2 + 1 = 0.$$

The origin is a lacunary value of $f_n(z)$ on $D \setminus \{0\}$ and the value 1 is attained only twice on D by $f_n(z)$. By the generalysed form of Montel's normal family criterion ([56], p.70 or Theorem 5, appendix B) this implies that the sequence $(g_n(z))$ is a normal family in $D \setminus \{0\}$. Hence, going to a subsequence, we can assume that $g_n(z)$ converges uniformly on compact subsets of $D \setminus \{0\}$. There are only two possible cases: either $g_n(z)$ tends locally uniformly to infinity (Case 1), or $g_n(z)$ tends locally uniformly to an analytic function in $D \setminus \{0\}$ (Case 2). We show in what follows that both these cases lead to a contradiction.

Case 1. $g_n(z)$ tends locally uniformly to infinity, i.e. the functions $\frac{1}{g_n(z)}$ tend locally to zero on every compact set of $D \setminus \{0\}$. Consider the compact set $\{z : |z| = \frac{1}{2}\}$. Given $\epsilon > 0$, we have $|\frac{1}{g_n(z)}| < \frac{\epsilon}{2} = \epsilon |z|$ for every $n \geq n_0(\epsilon)$ and $|z| = \frac{1}{2}$. By definition of $g_n(z)$ we know that

$$z f_n(z)\left(1 - \frac{1}{g_n(z)}\right) = -\left(1 + \frac{nz^2}{g_n(z)}\right).$$

And thus, using the bound obtained above, we get that, when $|z| = \frac{1}{2}$ and $n \geq n_0(\epsilon)$,

$$\frac{1}{2}|f_n(z)|(1 - \frac{\epsilon}{2}) < 1 + \frac{n\epsilon}{8}.$$

When $\epsilon < 2$ this implies then also

$$|\frac{f_n(z)}{n}| < 2\frac{\frac{1}{n} + \frac{\epsilon}{8}}{1 - \frac{\epsilon}{2}}.$$

When $n \geq n_0(\frac{1}{2})$ we have then $\epsilon < \frac{1}{2}$ and

$$|\frac{f_n(z)}{n}| < 2\left(\frac{\frac{1}{n} + \frac{1}{16}}{1 - \frac{1}{4}}\right) = \frac{1}{6} + \frac{8}{3n}.$$

For some large integer n we thus have

$$|\frac{f_n(z)}{n}| < \frac{1}{2}, \quad |z| = \frac{1}{2}$$

but also

$$|\frac{f_n(z)}{n}| < |z|, \quad |z| = \frac{1}{2}.$$

The functions $\frac{f_n(z)}{n}$ are analytic in $\{z : |z| \leq \frac{1}{2}\}$ and have a modulus on $\{z : |z| = \frac{1}{2}\}$ that is less than the modulus of the function $h(z) = z$. Hence, by Rouché's theorem, the function

$$\frac{f_n(z)}{n} - z$$

has a zero in $\{z : |z| \leq \frac{1}{2}\}$ for some integer n. This contradicts the fact that $f_n(z) - nz \in U(\overline{D})$ and thus case 1 can not occur.

Case 2. $g_n(z)$ tends locally uniformly to an analytic function in $D \setminus \{0\}$. Then $(g_n(z))$ are uniformly bounded on compact subsets of $D \setminus \{0\}$. Say that $|g_n(z)| \leq M$ for $|z| = \frac{1}{2}$. We have defined

$$g_n(z) = \frac{z(f_n(z) - nz)}{z f_n(z) + 1},$$

and thus also

$$g_n(z) = 1 - \frac{1 + nz^2}{zf_n(z) + 1}.$$

This last equation, together with the bound on $g_n(z)$, implies that

$$\left|\frac{1 + nz^2}{f_n(z)nz + 1}\right| \le M + 1, \ |z| = \frac{1}{2}.$$

This in turn implies that

$$\left|\frac{1}{f_n(z)nz + 1}\right| \le \frac{M + 1}{\frac{n}{4} - 1}, \ |z| = \frac{1}{2}, \ n > 4.$$

The function $\frac{1}{f_n(z)nz+1}$ is analytic in D and hence, by the Maximum Modulus Theorem, the bound obtained above for the function $\frac{1}{f_n(z)nz+1}$ holds throughout the disc of radius $\frac{1}{2}$. In particular it holds at $z = 0$ so that we must have

$$1 \le \frac{M + 1}{\frac{n}{4} - 1}, \ n > 4.$$

But this inequality is obviously violated when $n > 4(M + 2)$. A contradiction is obtained and thus Case 2 can not occur. ∎

We can give a slightly stronger formulation of Theorem 5.9 (wich we state here in a $\overline{D}$-stability context).

Theorem 5.10 *There exists a positive constant $\overline{\epsilon}$ such that the system*

$$p_\epsilon(z) = \frac{z}{z^2 + \epsilon}$$

is bistably $\overline{D}$-stabilizable when $\epsilon > \overline{\epsilon}$ or when $\epsilon = 0$ but that is not bistably $\overline{D}$-stabilizable when $0 < \epsilon < \overline{\epsilon}$.

Proof The proof follows from

1. the definition

$$\bar{\epsilon} = \sup_{\epsilon>0}\{\epsilon : \epsilon \text{ is such that } p_\epsilon(z) \text{ is not bistably stabilizable}\},$$

2. the fact, proved in Theorem 5.9, that $\bar{\epsilon}$ is thereby well defined,

3. the observation that, if $p_{\epsilon_0}(z)$ is not bistably stabilizable and $0 < \epsilon_1 < \epsilon_0$, then $p_{\epsilon_1}(z)$ is not bistably stabilizable either.

It is the last observation that we prove hereafter, the first point is a definition and the second one is already proved in Theorem 5.8.

Suppose, by contradiction, that $p_{\epsilon_0}(z)$ is not bistably stabilizable, that $0 < \epsilon_1 < \epsilon_0$ and that $p_{\epsilon_1}(z)$ is stabilizable by a bistable controller $c_1(z)$. We show that, from $c_1(z)$, we can construct a bistable controller $c_0(z)$ that stabilizes $p_0(z)$ and thus contradicts our assertion.

Note therefore that, because $c_1(z)$ stabilizes $p_{\epsilon_1} = \frac{z}{z^2+\epsilon_1}$, we have

$$zc_1(z) + z^2 + \epsilon_1 \in U(\overline{D})$$

and thus

$$zc_1(z) + z^2 + \epsilon_1 \neq 0, \ |z| \leq 1.$$

But then also, multiplying both sides by $\frac{\epsilon_0}{\epsilon_1}$

$$\frac{\epsilon_0}{\epsilon_1}zc_1(z) + \frac{\epsilon_0}{\epsilon_1}z^2 + \epsilon_0 \neq 0, \ |z| \leq 1.$$

This can be rewritten as

$$\sqrt{\frac{\epsilon_0}{\epsilon_1}}\sqrt{\frac{\epsilon_0}{\epsilon_1}}zc_1(\sqrt{\frac{\epsilon_1}{\epsilon_0}}\sqrt{\frac{\epsilon_0}{\epsilon_1}}z) + (\sqrt{\frac{\epsilon_0}{\epsilon_1}}z)^2 + \epsilon_0 \neq 0, \ |z| \leq 1.$$

By a change of variable, we get

$$\sqrt{\frac{\epsilon_0}{\epsilon_1}}zc_1(\sqrt{\frac{\epsilon_1}{\epsilon_0}}z) + z^2 + \epsilon_0 \neq 0, \ |z| \leq \sqrt{\frac{\epsilon_0}{\epsilon_1}}.$$

And, because $\sqrt{\frac{\epsilon_0}{\epsilon_1}} > 1$, we have then

$$\sqrt{\frac{\epsilon_0}{\epsilon_1}}zc_1(\sqrt{\frac{\epsilon_1}{\epsilon_0}}z) + z^2 + \epsilon_0 \neq 0, \ |z| \leq 1.$$

And thus the controller $c_0(z)$ defined by

$$c_0(z) \triangleq \sqrt{\frac{\epsilon_0}{\epsilon_1}}c_1(\sqrt{\frac{\epsilon_1}{\epsilon_0}}z)$$

is a stabilizing controller of $p_{\epsilon_0}(z) = \frac{z}{z^2+\epsilon_0}$. But since $c_1(\sqrt{\frac{\epsilon_1}{\epsilon_0}}z) \neq 0$ for all $z \in \overline{D}$, this means that $c_0(z)$ is bistable, a contradiction is achieved and the theorem is proved. ∎

The value of $\overline{\epsilon}$ is unknown. It was recently (1992 and 1993) shown that

$$\overline{\epsilon} < \frac{1}{e^2} = 0.1353...$$

and that

$$0.00000000036 < \overline{\epsilon}.$$

See Rupp ([101], 1993) for a proof of the existence of this lower bound. The two bounds above are, yet, the best bounds achieved.

5.4 Sufficient conditions for more than two systems

A trivial situation for which k systems are simultaneously stabilizable is the case where the systems are all stable. If $p_i \in \mathbf{R}(s)$ ($i = 1, ..., k$) are all stable then

$$\epsilon_i \triangleq \sup_{s \in \mathbf{C}_{+\infty}} |p_i(s)| \ (i = 1, ..., k)$$

are all well defined (finite) real numbers and any controller $c \in \mathbf{R}(s)$ that satisfies

$$\inf_{s \in \mathbf{C}_{+\infty}} |\frac{1}{c(s)}| = \frac{1}{\sup_{s \in \mathbf{C}_{+\infty}} |c(s)|} > \epsilon_i \ (i = 1, ...k)$$

is a simultaneous stabilizing controller for p_i $(i = 1, ..., k)$. Indeed, under these last conditions, the rational function $-\frac{1}{c(s)}$ avoids $p_i(s)$ in $\mathbf{C}_{+\infty}$ since the range of $-\frac{1}{c(s)}$ on $\mathbf{C}_{+\infty}$ is disjoint from that of $p_i(s)$.

More trivially, the controller $c = 0$ is a stabilizing controller for stable systems. In fact, any sufficiently small controller will be stabilizing for p_i. In particular, if a given rational function $q(s)$ is stable then it is always possible to choose a sufficiently small gain α for which the controller defined by $c(s) = \alpha q(s)$ fulfills the condition above. The same line of argument applies to inverstable systems which can always be simultaneously stabilized by an inverstable controller of which the poles and zeros are fixed and of which we tune the gain to a sufficiently large value.

The issue is somewhat different for systems that are minimum phase. A minimum phase rational function has no unstable zero except perhaps for a zero at infinity. Therefore the range of a minimum phase rational function on $\mathbf{C}_{+\infty}$ may include the origin and the above arguments do not apply to this case. However, because the only point to be mapped to 0 is the point at infinity, the range of a minimum phase function does not cover a ball centered at the origin. This is essentially what is proved in the next theorem for which we need a preliminary definition.

Definition 5.1 (High frequency gain) *Let p be a rational function. The high frequency gain of p is equal to the ratio between the coefficients of the highest terms in the numerator and of the denominator of p.*

Theorem 5.11 (Wei [116]) *Let p be a minimum phase rational function with positive (respectively negative) high frequency gain. Then there exists $\delta^* < 0$ (respectively $\delta^* > 0$) such that $p(s) \neq \delta$ for all $s \in C_{+\infty}$ and $\delta^* < \delta < 0$ (respectively $0 < \delta < \delta^*$).*

Using this property of minimum phase rational functions the next theorem is easy to prove.

Theorem 5.12 *Minimum phase systems that have the same high frequency gain sign are simultaneously stabilizable.*

Proof Let p_i $(i = 1, ..., k)$ be k minimum phase systems that have same high frequency gain sign. Assume without loss of generality that the sign is positive. By Theorem 5.11 there exists strictly negative values δ_i^* that satisfy $p_i(s) \neq \delta_i$ for all $s \in C_{+\infty}$ and $\delta_i^* < \delta_i < 0$. Define $\delta_{min} = \min_{i=1,...,k} \delta_i$ and observe that any value δ with $\delta_{min} < \delta < 0$ avoids p_i $(i = 1, ..., k)$ in $C_{+\infty}$. By Theorem 2.12 any controller of the form $c = -\frac{1}{\delta}$ with $\delta_{min} < \delta < 0$ is stabilizing for p_i $(i = 1, ..., k)$. ∎

We describe hereafter one more situation for which several systems are simultaneously stabilizable. The underlying idea is the following: k systems are simultaneously stabilizable if and only if there exists a rational function that avoids all of them in $C_{+\infty}$. Suppose now that among the k systems $p_1, ..., p_k$ is one system (say p_1) that avoids $p_2, p_2, ..., p_k$ in $C_{+\infty}$. Then by Theorem 2.12 $p_2, p_3, ..., p_k$ are stabilized by $c = -\frac{1}{p_1}$. In fact it is then possible to say more: if one system of the set $\{p_1, p_2, ..., p_k\}$ avoids all the other systems in $C_{+\infty}$ then $p_1, p_2, ..., p_k$ are simultaneously stabilizable.

Theorem 5.13 *Let $p_i \in R(s)$ $(i = 1, ..., k)$ and suppose that there exists a j $(1 \leq j \leq k)$ such that p_j avoids p_i in $C_{+\infty}$ $(i = 1, ..., k$ and $i \neq j)$. Then the systems p_i $(i = 1, ..., k)$ are simultaneously stabilizable.*

Proof Suppose without loss of generality that $j = 1$ and let $\frac{n_i}{d_i}$ be fractional factorizations of p_i in S. There exist $x, y \in S$ such that $n_1 x + d_1 y = 1$. Since p_1 avoids p_i in $C_{+\infty}$ $(i = 2, ..., k)$ we have $u_i \triangleq n_i d_1 - d_i n_1 \in U$ $(i = 2, ..., k)$. Therefore $\delta_{min} \triangleq$ $\min_{i=2,...,k} \frac{\inf_{s \in C_{+\infty}} |u_i(s)|}{\sup_{s \in C_{+\infty}} |x(s)n_i(s)+y(s)d_i(s)|}$ is well defined (finite) and strictly greater than zero. We choose δ with $0 < \delta < \delta_{min}$ and claim that $q \triangleq \frac{n_1 - \delta y}{d_1 + \delta x} \in R(s)$ avoids p_i $(i = 1, ..., k)$ in $C_{+\infty}$. ∎

Let $p_1 = \frac{1}{s-1}, p_2 = -\frac{s}{3s+1}, p_3 = -\frac{s-2}{5s-1}$ and $p_4 = -\frac{s^2-3s+1}{7s^2-s+2}$. p_1 intersects p_i $(i = 2, 3, 4)$ at the unique point -1 and hence the systems p_1, p_2, p_3 and p_4 are simultaneously stabilizable. For example $c = \frac{3}{2}$ is a simultaneous stabilizing controller.

5.5   Sufficient conditions from H_∞

As explained in the introductory chapter of this monograph the simultaneous stabilization question can be seen as one that arises in the context of robust control. The idea of robust control design is to obtain a controller that achieves stability even in the presence of uncertainties on a system. These uncertainties can be described in several different ways and our simultaneous stabilization approach corresponds to a 'finite approach' to robust control since we look only at a finite number of systems.

The H_∞ approach to robust control deals with infinitely many systems that all lie in a neighborhood of a nominal system and this philosophy is therefore more appropriate to deal with systems that are perturbed but remain resonably close to some nominal system.

In spite of this conceptual difference, results from H_∞ theory can be used to derive conservative sufficient conditions for simultaneous stabilization. We give hereafter an example of such an application. The H_∞ norm in S is the supremum norm i.e. if $p \in S$ then $||p||_\infty = \sup_{s\in\mathbf{C}_{+\infty}} |p(s)|$.

Theorem 5.14 (Additive perturbations) *Let $c \in \mathbf{R}(s)$ be a stabilizing controller of the system $p \in \mathbf{R}(s)$. If $||\frac{c}{1+pc}||_\infty \leq \epsilon$ then c stabilizes $p + \Delta_p$ for all $\Delta_p \in S$ with $||\Delta_p||_\infty < \frac{1}{\epsilon}$.*[1]

Proof Let $\frac{n_p}{d_p}$ be a fractional factorization of p in S and define $x, y \in S$ by $n_1 x + d_1 y = 1$. The stabilizing controller c is of the form $\frac{n_c}{d_c}$ with $n_c = x + rd_p$ and $d_c = y - rn_p$ for some $r \in S$. For this controller we have $\frac{c}{1+pc} = d_p n_c$. If $||\Delta_p||_\infty < \frac{1}{\epsilon}$ and $||\frac{c}{1+pc}|| \leq \epsilon$ then $||\Delta_p||_\infty < \frac{1}{||d_p n_c||_\infty}$ and thus $||\Delta_p d_p n_c||_\infty < 1$. But then $1 + \Delta_p d_p n_c$ is stable and never equal to zero in $\mathbf{C}_{+\infty}$ i.e. $1 + \Delta_p d_p n_c = u$ belongs to U. Since $n_p n_c + d_p d_c = 1$ this implies also that $1 + \Delta_p d_p n_c = n_p n_c + d_p d_c + \Delta_p d_p n_c = (n_p + \Delta_p d_p) n_c + d_p d_c = u \in U$. Therefore c is a stabilizing controller of $\frac{n_p + \Delta_p d_p}{d_p} = p + \Delta_p$ as requested. ∎

A direct application of this result is:

---

[1] Note that, since c is stabilizing, $\frac{c}{1+pc}$ is stable and, by the Maximum Modulus Theorem, $||\frac{c}{1+pc}||_\infty = \sup_{w\in\mathbf{R}} |\frac{c(iw)}{1+p(iw)c(iw)}|$.

Theorem 5.15 *Let $p_i \in \mathbf{R}(s)$ $(i = 1, ..., k)$ and assume that $p \in \mathbf{R}(s)$ and $c \in \mathbf{R}(s)$ satisfy the following three conditions:*

1. *c stabilizes p,*

2. *$p_i - p$ are stable $(i = 1, ..., k)$,*

3. *$\|p_i - p\|_\infty < \frac{1}{\|\frac{c}{1+pc}\|_\infty}$.*

Then c is a simultaneous stabilizing controller for the systems p_i $(i = 1, ..., k)$.

This is only one example of how one can apply a result of H_∞ robust control to obtain sufficient conditions for simultaneous stabilization. The array of results in H_∞ control theory is large. Results exist for different types of perturbation descriptions (multiplicative, additive, uncertainty in coprime factors, weighted perturbations, etc) and each of these results can be used to obtain sufficient conditions for simultaneous stabilization. In the concluding section we refer the reader to some of the extended literature on H_∞.

5.6 Summary and bibliography

The necessary stabilization condition found in the previous chapter are sufficient for the case of two systems but not for the case of three or more systems. This is essentially what is contained in the first two sections. The message of these sections is: conditions on the real positive axis do not suffice for three or more systems.

One of the merit of this negative result is that it sketches the connection that exists between simultaneous stabilization and properties of the range of analytic functions. This connection is the key to any deep study of simultaneous stabilization of linear systems.

The sufficient conditions given in Section 4 and Section 5 are important from a practical point of view. Real life systems are generically minimum phase and so Theorem 5.12 may be used for these practical situations. From a theoretical point of view however, these results bypass the genuine problem and leave it untouched. We also point to the fact that even for minimum phase systems the same high frequency gain sign condition is only a necessary condition. There exist yet no tractable necessary and sufficient conditions under which minimum phase systems are simultaneously stabilizable.

Historically the first proof of Theorem 5.2 was given in Youla [123]. The proof given in this reference is constructive but has the disadvantage of being rather tedious. We have prefered the elegant version of Vidyasagar [108].

Strong stabilization and simultaneous stabilization of two systems ultimately rely on interpolation properties by bistable rational functions. Several other questions arising in system theory may be formulated in terms of interpolation, see Ghosh [52], Khargonekar and Tannenbaum [74] or Helton [61]. However, contrary to what may appear from the literature on the subject, simultaneous stabilization of three or more systems has not yet been formulated in terms of interpolation only.

Concerning the degree of a stable controller, interested readers may wish to look at Dorato [36], Smith [104] and Ohta [95] for a first introduction to this problem. The fact that the degree of a stable stabilizing controller is not a-priori bounded by the degree of the system can be found for example in Ghosh [50]. A proof involving results from analytic function theory that may help towards finding the lowest possible degree is contained in Blondel [21].

The results of the third section are contained in Blondel *et al.* [16, 19]. Theorem 5.9 disproves a conjecture made by Dr K. Wei in 1988 at the IFAC working group on robust control theory in Jeruzalem. The same conjecture appears explicitly in Wei [118,119]. References for the three key results of Section 3 ($\frac{1}{16}$-Theorem, Bermant's theorem and Montel's generalized normal family criterion) are given in Appendix C.

The first sufficient conditions of Section 4 are contained in Kwakernaak [81] and several successive refinements for multivariable systems, for infinitely many systems or for instability regions that are different from the extended right half plane can be found in Wei [116,117].

Theorem 5.14 can be found in any elementary introduction on H_∞ control design. See for example the books of Francis [44] or McFarlane and Glover [88]. An original approach (the polynomial approach) to H_∞ control is given in several papers of Kwakernaak ([82] and references therein). These references also contain efficient algorithms and constructive procedures for H_∞ control design.

Chapter 6

Necessary and sufficient conditions: rational decidability

Une branche de la science est vivante aussi longtemps qu'elle offre une foule de problèmes. Le manque de problèmes signifie sa mort.

J. Dieudonné, Pour l'honneur de l'esprit humain, 1987.

6.1 Introduction

Before explaining the contribution of this chapter we briefly summarize the main results of the previous chapters on simultaneous stabilization of k systems.

The case $k = 1$ –the stabilization of a single system– is dealt with in Chapter 3: there always exists a stabilizing controller for a single system. Moreover, once a stabilizing controller is found, the Youla-Kucera parametrization provides the set of all stabilizing controllers.

By using this parametrization, it is possible to rephraze simultaneous stabilization of two systems into strong stabilization of a single system. The strong stabilization question has an elegant solution: a system is stabilizable by a stable controller if and only if it has an even number of real unstable zeros between each pair of real unstable poles. This condition can be checked by performing only rational operations (additions, substractions, multiplications and divisions) on the coefficients of the system. The real unstable poles and zeros do not have to be computed explicitly.

Thus the simultaneous stabilizability question for two systems is fully solved; we can first translate it into a strong stabilization question by using the Youla-Kucera parametrization and then check the parity interlacing property.

For three or more systems the picture is different. We have given equivalent formulations (Chapter 3), necessary conditions (Chapter 4) or sufficient conditions (Chapter 5) but no tractable necessary and sufficient conditions.

Our ambition in this chapter is to show that no such tractable necessary and sufficient condition exist for simultaneous stabilization of three or more systems. The simultaneous stabilizability question for three or more systems is rationally undecidable: it is not possible to find a general criterion that involves only the coefficients of three or more linear systems, rational operations, logical operations ('and' and 'or') and sign tests operations (equal to, greater than, greater than or equal to, etc) and that is necessary and sufficient for simultaneous stabilizability of the systems.

Questions that can be solved by using only the above mentioned elementary operations are sometimes simply refered to as tractable. This chapter points thus to an intrinsic limitation of simultaneous stabilizability conditions: it is possible to find tractable necessary *or* tractable sufficient conditions for simultaneous stabilization of three systems but it is not possible to find tractable necessary *and* sufficient conditions.

The notion of rational decidability is presented in Section 2. In Section 3 we show in Theorem 6.4 that simultaneous stabilizabition of three or more linear systems is not rationally decidable.

6.2 Rational decidability and algebraic numbers

A question is rationally decidable if it can be answered by using only elementary operations. We formalize this concept with the next two definitions.

Definition 6.1 (Elementary operations) *An elementary operation is any one of*

1. *the four algebraic operations: addition, substraction, multipli-cation and division,*

2. *the two logical operations: 'and' and 'or',*

3. *the five test operations: $=, >, <, \geq$ or $\leq$.*

Definition 6.2 *A binary question $Q(a_1, ..., a_n)$ associated to an n-uple $(a_1, ..., a_n) \in \mathbf{R}^n$ is rationally decidable if and only if there exists a logical sentence $L(a_1, ..., a_n)$ of finite length that involves only elementary operations on the entries a_i of the n-uple and that is such that*

$$\text{for all } (a_1, ..., a_n) \in \mathbf{R}^n \ (L \text{ is true if and only if } Q \text{ is}).$$

The following binary questions are all rationally decidable: testing the stability of a polynomial, the positive definiteness of a matrix, the coprimeness of two polynomials or the simultaneous stabiliz-ability of two linear systems.

Rationally decidable questions are closely related to algebraic numbers.

Definition 6.3 *A real number is algebraic if and only it is the root of a polynomial that has integer (or rational) coefficients. A real number that is not algebraic is transcendental.*

$-1, \sqrt{2}, i = \sqrt{-1}$ and $\frac{\sqrt{\sqrt{7}+31}}{\sqrt{13}-5}$ are examples of algebraic numbers. $\pi, e, 2^{\sqrt{2}}$ and $\Gamma(\frac{1}{4})$ are transcendental. There exists no systematic procedure for proving that a number is transcendental or not. Some simple numbers are still not proved to be transcendental or not[1].

---

[1] A systematic investigation of the irrationality and transcendence of real numbers was the 7th of Hilbert's 23 famous problems which he adressed at the second International Congress of Mathematics in Paris in 1900.

For our simultaneous stabilization purposes we need the next non-trivial result.

Theorem 6.1 *The real number $\frac{4\pi^2}{\Gamma^4(\frac{1}{4})}$ is transcendental.*

Proof The proof is based on a result contained in the third section of the last chapter of *Transcendental number theory* (Baker, p. 158, [10]). This result states: "The transcendence degree of the field L generated by $\omega_1 = \frac{\Gamma^2(\frac{1}{4})}{\sqrt{8\pi}}$, $\omega_2 = i\omega_1$, $\eta_1 = \frac{\pi}{\omega_1}$, and $\eta_2 = -i\eta_1$ over the rationals $\mathbf{Q}$ is at least 2."
Since $\omega_1^2 + \omega_2^2 = 0$ and $\eta_1^2 + \eta_2^2 = 0$ this means that ω_1 and η_1 are both transcendental and are algebraically independent. Then $\frac{\eta_1}{2\omega_1^3} = \frac{4\pi^2}{\Gamma^4(\frac{1}{4})}$ is transcendental and so the theorem is proved. ∎

A connection between algebraic numbers and rationally decidable questions is as follows. Remember that $\mathbf{Q}(\beta)$ denotes the set of rational functions in the variable β and with rational coefficients.

Theorem 6.2 *Let $Q(a_1, ..., a_n)$ be a rationally decidable question associated to an n-uple $(a_1, ..., a_n)$. If all the entries a_i of the n-uple are in $\mathbf{Q}(\beta)$ then $Q(a_1(\beta), ..., a_n(\beta))$ is true if and only if β belongs to a finite union of closed-open intervals whose endpoints are algebraic numbers or are infinite.*

Proof Since the question $Q(a_1, ..., a_n)$ is decidable, there exists a logical sentence $L(a_1, ..., a_n)$ of finite length that involves only elementary operations on the entries a_i of the n-uple and that is such that $L(a_1, ..., a_n)$ is true if and only if $Q(a_1, ..., a_n)$ is. Thus, for all real β

$$Q(a_1(\beta), ..., a_n(\beta)) \text{ is true} \Leftrightarrow L(a_1(\beta), ..., a_n(\beta)) \text{ is true.}$$

It remains to show that there exist values $\overline{\sigma}_{k,j}$ and $\underline{\sigma}_{k,j}$ ($k = 1,2$ and $j = 1,...,m_k$) that are either equal to $\pm\infty$ or to algebraic numbers, such that

$$\forall\beta \in \mathbf{R} : L(a_1(\beta),...,a_n(\beta)) \text{ is true} \Leftrightarrow$$

$$\beta \in \left(\cup_{j=1}^{m_1}(\underline{\sigma}_{1,j},\overline{\sigma}_{1,j}]\right) \cup \left(\cup_{j=1}^{m_2}[\underline{\sigma}_{2,j},\overline{\sigma}_{2,j})\right).$$

To prove this we proceed by induction on the size of the logical sentence $L(a_1,...,a_n)$.

The logical sentence $L(a_1,...,a_n)$ is either made up of two smaller logical sentences $L_1(a_1,...,a_n)$ and $L_2(a_1,...,a_n)$ linked by an 'and' (denoted by $\wedge$) or an 'or' (denoted by $\vee$) logical operation ($L(a_1, ...,a_n) = L_1(a_1,...,a_n) \wedge L_2(a_1,...,a_n)$ or $L(a_1, ...,a_n) = L_1(a_1, ...,a_n) \vee L_2(a_1,...,a_n)$) or is a nucleus expression of the form $L(a_1, ...,a_n) = R_1(a_1,...,a_n) \square R_2(a_1,...,a_n)$ where $R_1(a_1,...,a_n)$ and $R_2(a_1,...,a_n)$ are rational expressions of the coefficients $a_1,...,a_n$ and $\square$ is any one of the five sign test operations $<,\leq,>,\geq,=$.

We analyse these two cases successively.

First, if $L(a_1,...,a_n)$ is a nucleus expression, then $L(a_1(\beta),...,a_n(\beta))$ is true if and only if

$$R_1(a_1(\beta),...,a_n(\beta)) \square R_2(a_1(\beta),...,a_n(\beta))$$

for some $\square \in \{<,\leq,>,\geq,=\}$. By hypothesis $a_i(\beta)$ are rational expressions of β and $R_j(a_1,...,a_n)$ are rational expressions of $a_1,...,a_n$. Hence the functions of β defined by $R'_j(\beta) = R_j(a_1(\beta),...,a_n(\beta))$ are also rational expressions of β. The condition $R_1(a_1(\beta),...,a_n(\beta)) \square R_2(a_1(\beta),...,a_n(\beta))$ is satisfied if and only if $R'_1(\beta) \square R'_2(\beta)$ is, and this last condition is equivalent to

$$\beta \in \left(\cup_{j=1}^{m_1}(\underline{\sigma}_{1,j},\overline{\sigma}_{1,j}]\right) \cup \left(\cup_{j=1}^{m_2}[\underline{\sigma}_{2,j},\overline{\sigma}_{2,j})\right)$$

for some $\overline{\sigma}_{k,j}$ and $\underline{\sigma}_{k,j}$ ($k = 1, 2$ and $j = 1, ..., m_k$) that are equal to $\pm\infty$ or to algebraic numbers. Thus the theorem is proved in the case of a nucleus expression.

Secondly, suppose that $L(a_1, ..., a_n)$ is made up of two logical sentences

$L_1(a_1, ..., a_n)$ and $L_2(a_1, ..., a_n)$ linked by an 'and' or an 'or' logical operation.

By the induction hypothesis assume that the values $\overline{\sigma}_{k,j}^1$ and $\underline{\sigma}_{k,j}^1$ ($k = 1$, 2 and $j = 1, ..., m_k^1$) and $\overline{\sigma}_{k,j}^2$ and $\underline{\sigma}_{k,j}^2$ ($k = 1, 2$ and $j = 1, ..., m_k^2$) are equal to $\pm\infty$ or to algebraic numbers and are such that

$$L_1(a_1(\beta), ..., a_n(\beta)) \text{ is true} \Leftrightarrow$$

$$\beta \in \left(\cup_{j=1}^{m_1}(\underline{\sigma}_{1,j}^1, \overline{\sigma}_{1,j}^1]\right) \cup \left(\cup_{j=1}^{m_1}[\underline{\sigma}_{2,j}^1, \overline{\sigma}_{2,j}^1)\right)$$

and

$$L_2(a_1(\beta), ..., a_n(\beta)) \text{ is true} \Leftrightarrow$$

$$\beta \in \left(\cup_{j=1}^{m_1}(\underline{\sigma}_{1,j}^2, \overline{\sigma}_{1,j}^2]\right) \cup \left(\cup_{j=1}^{m_2}[\underline{\sigma}_{2,j}^2, \overline{\sigma}_{2,j}^2)\right).$$

Then, if $L(a_1, ..., a_n) = L_1(a_1, ..., a_n) \wedge L_2(a_1, ..., a_n)$, we have

$$L(a_1(\beta), ..., a_n(\beta)) \text{ is true} \Leftrightarrow$$

$$\beta \in \left(\left(\cup_{j=1}^{m_1}(\underline{\sigma}_{1,j}^1, \overline{\sigma}_{1,j}^1]\right) \cup \left(\cup_{j=1}^{m_2}[\underline{\sigma}_{2,j}^1, \overline{\sigma}_{2,j}^1)\right)\right)$$
$$\cap \left(\left(\cup_{j=1}^{m_1}(\underline{\sigma}_{1,j}^2, \overline{\sigma}_{1,j}^2]\right) \cup \left(\cup_{j=1}^{m_2}[\underline{\sigma}_{2,j}^2, \overline{\sigma}_{2,j}^2)\right)\right).$$

On the other hand, if $L(a_1, ..., a_n) = L_1(a_1, ..., a_n) \vee L_2(a_1, ..., a_n)$, we have

$$L(a_1(\beta), ..., a_n(\beta)) \text{ is true} \Leftrightarrow$$

$$\beta \in \left(\left(\cup_{j=1}^{m_1}(\underline{\sigma}_{1,j}^1, \overline{\sigma}_{1,j}^1]\right) \cup \left(\cup_{j=1}^{m_2}[\underline{\sigma}_{2,j}^1, \overline{\sigma}_{2,j}^1)\right)\right)$$
$$\cup \left(\left(\cup_{j=1}^{m_1}(\underline{\sigma}_{1,j}^2, \overline{\sigma}_{1,j}^2]\right) \cup \left(\cup_{j=1}^{m_2}[\underline{\sigma}_{2,j}^2, \overline{\sigma}_{2,j}^2)\right)\right).$$

In both cases we can rewrite the unions and intersections involved under the form

$$\left(\cup_{j=1}^{m_1}(\underline{\sigma}_{1,j}, \overline{\sigma}_{1,j}]\right) \cup \left(\cup_{j=1}^{m_2}[\underline{\sigma}_{2,j}, \overline{\sigma}_{2,j})\right)$$

for some $\overline{\sigma}_{k,j}$ and $\underline{\sigma}_{k,j}$ ($k = 1, 2$ and $j = 1, ..., m_k$) equal to $\pm\infty$ or to algebraic numbers. Thus, by induction on the size of L, the theorem is proved. ∎

6.3   Simultaneous stabilization of 3 or more systems

We first need to introduce a result from analytic function theory. In all what follows we define $A \triangleq \frac{4\pi^2}{\Gamma^4(\frac{1}{4})} = 0.2284732...$

Theorem 6.3 *Let $\beta \in \mathbf{R}$. There exists a real function $f(z)$ that is analytic on D, such that $f(0) = 0$, $f'(0) = 1$ and that leaves out the values β and $-\beta$ on D if and only if $|\beta| \geq A$.*

Proof We first prove sufficiency. Let $f_e(z)$ be the real function defined by the infinite converging product

$$f_e(z) \triangleq A \left(32 e^{-\pi \frac{1+z}{1-z}} \prod_{n=1}^{\infty} \left(\frac{1 + e^{-2n\pi \frac{1+z}{1-z}}}{1 - e^{-(2n-1)\pi \frac{1+z}{1-z}}}\right) - 1\right).$$

It is shown in Nehari [93] (bottom of page 330) that $f_e(z)$ is analytic on $|z| < 1$, is such that $f_e(0) = 0$ and $f_e'(0) = 1$ and that $f_e(z) \neq A$ and $f_e(z) \neq -A$ when $z \in D$.

Assume now that $\beta \geq A$ and define

$$f(z) \triangleq \frac{\beta}{A} f_e(\frac{A}{\beta} z).$$

Due to the properties of $f_e(z)$ it is easy to check that $f(z)$ is analytic on $|z| < 1$ (note that this fails when $|\beta| < A$), is such that $f(0) = 0$ and $f'(0) = 1$ and that $f(z) \neq \beta$ and $f(z) \neq -\beta$ when $z \in D$. This ends the first part of the proof.

For necessity, assume by contradiction that $f(z)$ satifies the conditions of the theorem and that $0 < \beta < A$. By assumption, the image of the disc D under the mapping $\xi = f(z)$ contains neither the value β nor the value $-\beta$. Thus, the image does not cover any segment of the real line that contains the origin and is of length $2A$. This contradicts Theorem C.3 in Appendix C, a contradiction is obtained and the theorem is proved. ∎

With the help of this theorem we prove:

Theorem 6.4 *The simultaneous stabilizability of two second order systems by a stable controller is rationally undecidable.*

Proof Discrete and continuous time stability regions are mapped into each other by the usual bilinear transformation. Rational decidability in discrete time implies rational decidability in continuous time and vice-versa. Here we choose a discrete time set up for the proof.

Assume that $\beta \in \mathbf{R}$ and consider the two systems $p_{1,\beta}(z) = \frac{z^2}{z^2 - \beta}$ and $p_{2,\beta}(z) = \frac{z^2}{z^2 + \beta}$.

We proceed in two steps.

First, we show that when $\beta = 0$ or when $|\beta| > A$ the two systems are simultaneously stabilizable by a stable controller whereas when $0 < |\beta| < A$ they are *not* simultaneously stabilizable by a stable controller. (Note that we leave out the analysis of the case $\beta = A$.)
Second, we show that the first step contradicts the fact that the simultaneous stabilizability question of two systems by a stable con-

troller is rationally decidable.

Step 1. If $\beta = 0$ then $p_{1,\beta}(z) = p_{2,\beta}(z) = z$ and a stable stabilizing controller is given, for example, by $c(z) = 2$.

If $\beta \neq 0$ then the two systems are simultaneously stabilizable by a stable controller if and only if there exists a rational function $c(z)$ that has no poles in $\overline{D}$ and such that

$$z^2 c(z) + z - \beta$$

and

$$z^2 c(z) + z + \beta$$

have no zeros in $\overline{D}$.

It remains to show that, when $\beta < A$ such a function $c(z)$ does not exist whereas it does exist when $\beta > A$. We prove these two points in (a) and (b) respectively.

(a) Assume, by contradiction, that $|\beta| < A$, that $c(z)$ has no poles in $\overline{D}$ and that

$$z^2 c(z) + z - \beta$$

and

$$z^2 c(z) + z + \beta$$

have no zeros in $\overline{D}$. Then, the function defined by

$$f(z) \triangleq z^2 c(z) + z$$

satisfies all the conditions of Theorem 6.3 and leaves out the values $\pm\beta$ on D. $\beta < A$, a contradiction is thus achieved and this part is proved.

(b) Assume that $|\beta| > A$. We construct a rational function $c(z)$ that satisfies all the required conditions. By Theorem 6.3, there exists an analytic function $f(z)$ on D such that $f(0) = 0$, $f'(0) = 1$ and that leaves out the values $\pm A$. We define the function $g(z)$ by

$$g(z) \triangleq \frac{\beta}{A} f\left(\frac{A}{\beta} z\right).$$

Due to the properties of $f(z)$, the function $g(z)$ is such that

1. $g(\bar{z}) = \overline{g(z)}$,

2. $g(z)$ is analytic on $|z| < \frac{\beta}{A}$,

3. $g(0) = 0$ and $g'(0) = 1$,

4. $g(z)$ leaves out the values $\pm A$ on $|z| < \frac{\beta}{A}$.

Note that, in these expressions, $1 < \frac{\beta}{A}$ and so $g(z)$ is analytic on D. In the sequel we construct, with the help of this function $g(z)$, a real *polynomial* that shares the same properties.

Because of the properties 2 and 4, the real number μ defined by

$$\mu \triangleq \min\{\inf_{z \in \bar{D}} |g(z) - A|, \inf_{z \in \bar{D}} |g(z) + A|\}$$

is strictly positive. Because of the first three properties, the function $h(z)$ defined by

$$h(z) \triangleq \frac{g(z) - z}{z^2}$$

is real and analytic in $\{z : |z| < \frac{\beta}{A}\}$. By Runge's theorem (see Rudin [96]), there exists a real polynomial $q(z)$ such that

$$|h(z) - q(z)| < \mu \left(\frac{A}{\beta}\right)^2, \quad z \in \bar{D}.$$

This polynomial is then also such that

$$|g(z) - z - z^2 q(z)| < \mu, \ z \in \overline{D}.$$

Defining the polynomial $p(z)$

$$p(z) \triangleq z + z^2 q(z) \in \mathbf{R}[z]$$

we have $p(0) = 0$ and $p'(0) = 1$. Since

$$|g(z) - p(z)| < \mu, \ z \in \overline{D}$$

and

$$\mu \leq \min\{\inf_{z \in \overline{D}} |g(z) - A|, \inf_{z \in \overline{D}} |g(z) + A|\}$$

it follows that

$$|g(z) - p(z)| < |g(z) \pm A|, \ z \in \overline{D}.$$

Hence

$$p(z) \neq \pm A, \ z \in \overline{D}$$

as requested.

A polynomial is a rational function with no poles of module less than or equal to one and point (b) is thus proved.

Step 2. Assume, by contradiction, that the simultaneous stabilizability of two systems by a stable controller is a rationally decidable question. Then, so is the simultaneous stabilizability of the two systems $p_{1,\beta}(z) = \frac{z^2}{z^2 - \beta}$ and $p_{2,\beta}(z) = \frac{z^2}{z^2 + \beta}$ by a stable controller. But then, using Theorem 6.2, there exist values $\overline{\sigma}_{k,j}$ and $\underline{\sigma}_{k,j}$ ($k = 1, 2$ and $j = 1, ..., m_k$) that are either equal to $\pm\infty$ or to algebraic numbers, such that the three systems are simultaneously stabilizable if and only if

$$\beta \in \left(\cup_{j=1}^{m_1} (\underline{\sigma}_{1,j}, \overline{\sigma}_{1,j}] \right) \ \cup \ \left(\cup_{j=1}^{m_2} [\underline{\sigma}_{2,j}, \overline{\sigma}_{2,j}) \right).$$

But this contradicts our first step since we know from there that the two systems are simultaneously stabilizable by a stable controller if and only if

$$\beta \in (-\infty, -\frac{4\pi^2}{\Gamma^4(\frac{1}{4})}) \cup [0,0] \cup (\frac{4\pi^2}{\Gamma^4(\frac{1}{4})}, +\infty)$$

or if and only if

$$\beta \in (-\infty, -\frac{4\pi^2}{\Gamma^4(\frac{1}{4})}] \cup [0,0] \cup [\frac{4\pi^2}{\Gamma^4(\frac{1}{4})}, +\infty).$$

By Theorem 6.1 $\frac{4\pi^2}{\Gamma^4(\frac{1}{4})}$ is a transcendental number, a contradiction is achieved and the theorem is proved. ■

6.4 Summary and bibliography

In this chapter it has been shown that the simultaneous stabilization question for three or more systems is not rationally decidable. The notion of rational decidability is relatively limited in that it allows only the use of very elementary operations. It is not yet known if the problem would admit a solution in terms of finitely many transcendental functions in one or two arguments.

The main result of this chapter is to be found in Blondel and Gevers [22]. The terminology 'rational decidability' was suggested by Eduardo Sontag. The results of this chapter bear some resemblance with concepts associated to semi-algebraic sets and with the fundamental theory developed in several papers of Ghosh [52,53,54].

Epilogue

Tant qu'il y a de la vie
il y a de l'espoir.

Belgian proverb.

Catholic University of Louvain, Louvain-la-Neuve, Belgium, 1992.

S.D. Well?

V.B. ...interesting and tantalizing. Couldn't sleep some nights. 'Simultaneous' is a word that will probably sound strange to my ears 'till my death.

S.D. Time hasn't come yet. But never know, give me your conclusion before anything happens.

V.B. Didn't you read the monograph?

S.D. *Ahem... it's a good exercise for you; did you not quote some-*
where "Trying to express complex issues in very simple lan-
guage is an excellent exercise in discovering how well you re-
ally understand them"?

Try to summarize everything in a few points.

Say four points.

V.B. *Four points! a whole monograph in four points! You've got*
no heart. You're asking for a telegram not for a summary...

I'll take five points.

1. Stabilization = avoidance.

2. Simultaneous stabilization of k systems, simultaneous strong stabilization of $k - 1$ systems and
   simultaneous bistable stabilization of $k - 2$ systems; different words, same problems.

3. $R_{+\infty}$-stabilization conditions are necessary conditions for

   $C_{+\infty}$-stabilization; they lead to tractable interlacing conditions.

4. Necessary and sufficient conditions for $C_{+\infty}$-sta-
   bilization are hard to find. They are connected
   with questions in analytic function theory that
   are simple to state but as yet unsolved.

S.D. *Good. I'll think about it... but wait, did you not ask for five*
points instead of four?

V.B. *Yes. But I'm not sure...*

S.D. Go ahead.

V.B. Well, it's a kind of general conclusion. I don't know if you'll like it. To put it in one line:

5. Necessary and sufficient conditions for simultaneous stabilization of three or more systems do not exist.

S.D. Don't be ridiculous. What do you mean by "do not exist"?

V.B. It's a way of speaking. Necessary and sufficient conditions exist but what I mean is that they are useless because they are untractable. A good example of such useless answer is given by the 'equivalences' in simultaneous stabilization. For example:

Let p_1, p_2 and p_3 be three scalar linear systems and assume that p_1 is stable. Then the three systems are simultaneously stabilizable if and only if the two systems $p_2 - p_1$ and $p_3 - p_1$ are simultaneously stabilizable by a stable controller.

That's a necessary and sufficient condition for simultaneous stabilization.

S.D. Yes, but you have to assume that p_1 is stable.

V.B. No problem. You may forget this and give a similar formulation that doesn't need this assumption.

S.D. Allright then, but can you tell when two systems are simultaneously stabilizable by a stable controller?

V.B. No.

S.D. But then you haven't solved the question.

V.B. Yes I have. You wanted me to give a necessary and sufficient condition for simultaneous stabilization of three systems. That's what I've done. I have solved your question.

S.D. That's not really what I wanted. I wanted you to give me tractable necessary and sufficient conditions for simultaneous stabilization of three systems.

V.B. What do you mean by tractable?

S.D. Well... say conditions that you can actually implement. Conditions that involve only a finite number of operations.

V.B. O.K. Let me state my point 5. again:

5. Tractable necessary and sufficient conditions for simultaneous stabilization of three or more systems do not exist.

or, if you prefer:

5. It is not possible to find a general test that

- **uses the coefficients, poles and zeros of three systems**
- **performs rational (additions, substractions, multiplications and divisions) and logical operations**
- **gives a yes-no answer regarding the simultaneous stabilizability of the three systems in a finite number of steps.**

S.D. ...

V.B. What do you think?

S.D. Hu, hu... not too bad. I'm not sure that I believe this, how can you prove such a thing?... first of all, is it properly proved?

V.B. Yes. Proved and double checked. Everything seems to be O.K.

S.D. Let me see...

What about computing roots or evaluating exponentials and logarithms? and performing computations with π, e and $\Gamma(\frac{1}{3})$ or using trigonometric functions? are these not all tractable operations?

V.B. Not in the sense given above, but...

S.D. Listen Vincent, I'm sorry but we are running out of paper. I think that you should probably add one last concluding line before Springer-Verlag shuts the light off.

V.B. Right. But what? Any idea?

S.D. Aha.

V.B. What do you mean "Aha"? yes or no?

S.D. Give me your pen for a second: **6. Even after this result on untractable stuff, there is still plenty of room for fascinating and rewarding research on simultaneous stabilization to be done.**

V.B. Sounds fair, my turn, for the rewarding bit: **6bis. One kilogram of good world-famous Belgian chocolate ('pralines') to the first person who gives me a stable and minimum phase controller that stabilizes $\frac{(s-1)^2}{s^2-1.8s+1}$ or who convices me that such a controller doesn't exist.**

S.D. *And a second kilogram, not to be eaten the same day, for the first person to answer the more complete question: for what values of $\delta \in \mathbf{R}$ is the scalar linear system $p(s) = \frac{s^2-1}{s^2-2\delta s+1}$ stabilizable by a controller that is both stable and minimum phase?*

Appendix A

Rings and Algebras

We refer the interested readers to Jacobson [68], Cohn [27], Lang [84], MacDuffee [87] or Atiyah [9] for motivations of the definitions and theorems on rings. An elementary introduction to algebras is given in Chapter 18 of Rudin [96]. More advanced references include Duren [38] and Hoffman [64].

Throughout this appendix 'ring' really means 'commutative ring with identity' and is denoted by R.

Definition A.1   • *Assume that $x, y \in R$, $x \neq 0$. x divides y (or, that y is a multiple of x) if there exists $z \in R$ such that $y = xz$.*

- *Assume that $x \in R$, $x \neq 0$. x is a unit of R (or, is an invertible element of R) if x divides 1.*

- *A field is a ring in which every non zero element is invertible.*

- *Assume that $x, y \in R$. A greatest common divisor of x and y is any element $z \in R$ such that*

     1. z divides x and y,

2. *if t divides x and y then t also divides z.*

- *Assume that $x, y \in R$. x and y are relatively prime (or co-prime) if every greatest common divisor of x and y is invertible.*

- *A subset I of a ring R is an ideal of the ring R if*

    1. *$a \in I, b \in I \Rightarrow a + b \in I$,*

    2. *$(I, +)$ is a subgroup of $(R, +)$,*

    3. *$a \in I, x \in R \Rightarrow ax \in I$.*

- *Assume that $a \in R$. Then the set defined by $I \triangleq \{ax : x \in R\}$ is an ideal of R. It is called a principal ideal of R.*

- *A ring R is a principal ideal ring if all the ideals of R are principal.*

- *Assume that $x, y \in R$. R is a domain if*

$$xy = 0 \Rightarrow x = 0 \text{ or } y = 0.$$

- *A ring is a principal ideal domain if it is a domain and a principal ideal ring.*

- *A ring is an Euclidean ring if there is an integer valued function δ defined on $R \setminus \{0\}$ that satisfies the following two conditions:*

    1. *$x \in R$ divides $y \in R \Rightarrow \delta(x) \leq \delta(y)$,*

    2. *$\forall x, y \in R, y \neq 0$ there exists $q \in R$ such that either $x - qy = 0$ or $x - qy = r$ and $\delta(r) < \delta(y)$.*

- *A ring is an Euclidean domain if it is a domain and an Euclidean ring.*

Theorem A.1 *Euclidean domains are principal ideal domains.* ∎

Remark Principal ideal domains benefit from nice properties. They are, after the Euclidean rings, the rings that are, in a certain sense, the closest to the structure of a field. The next results are valid for rings R that are principal ideal domains.

Theorem A.2 *From R we construct a field F, the field of fractions of R, that contains R as a subring. The construction procedure is outlined hereafter. Consider $R \times R \backslash \{0\}$ and define an equivalence relation $\cong$ on $R \times R \backslash \{0\}$ by*

$$(a, b) \cong (c, d) \Leftrightarrow ad = bc.$$

$\cong$ is then an equivalence relation. The set of equivalence classes in $R \times R \backslash \{0\}$ is denoted by F. The set F consists of pairs (a, b) that can be thought of as fractions $\frac{a}{b}$. On F we define an addition and a multiplication by

$$(a, b) + (c, d) = (ad + bc, bd)$$

and

$$(a, b).(c, d) = (ac, bd).$$

These operations of addition and multiplication are well defined on F. Each member of R can be identified with the corresponding class of equivalence of $(a, 1)$ and every non-zero element of F is invertible since

$$(a, b).(b, a) \cong (1, 1).$$

F is referred to as the field of fractions of R. ∎

Theorem A.3 *Assume that R is a principal ideal domain. Let F be the field of fractions of R. Given any fraction $(a,b) \in F$ there exists an equivalent fraction $(c,d) \in F$ (i.e. $(a,b) \cong (c,d)$) such that c and d are coprime in R. Such an equivalent fraction is called a coprime fractional factorization of the fraction (a,b).* ∎

Theorem A.4 *Assume that R is a principal ideal domain. Then every pair $a,b \in R$ with at least one of a or b non-zero have a greatest common divisor $c \in R$ which can be expressed in the form $ax + by = c$ for some $x,y \in R$. Moreover, the set of all greatest common divisors of a and b is given by $\{cu, u$ is an invertible element of $R\}$.* ∎

Definition A.2
- $R^{n \times m}$ *denotes the set of $n \times m$ matrices whose entries are in the ring R.*

- *Sum, products, determinants, adjoints, minors, row and column are defined as usual.*

- *A square matrix $A \in R^{n \times n}$ is unimodular if it is invertible, i.e. if there exists a matrix denoted by A^{-1} in $R^{n \times n}$ such that $AA^{-1} = I_n$.*

- *The normal rank of a matrix $A \in S^{n \times m}$ is the largest integer l such that*
  1. *there exists a non-zero minor of dimension l,*
  2. *all the minors of dimension $(l+1) \times (l+1)$ are identically equal to zero.*

Theorem A.5 *Assume that $A \in R^{n \times m}$ and let r be the normal rank of A. Then there exist two unimodular matrices $U \in R^{n \times n}$ and $V \in R^{m \times m}$ such that $UAV = \Lambda$ and*

$$\bullet \ \Lambda(s) = \begin{pmatrix} \lambda_1(s) & & & & & & 0 \\ & \lambda_2(s) & & & & & \\ & & \ddots & & & & \\ & & & \lambda_r(s) & & & \\ & & & & 0 & & \\ & & & & & \ddots & \\ 0 & & & & & & 0 \end{pmatrix}$$

- $\lambda_i(s)$ divides $\lambda_{i+1}(s)$ $(i = 1, \ldots, r-1)$.

∎

Definition A.3 • *An algebra A over a field F $(= \mathbf{R}, \mathbf{C})$ is a pair consisting of a ring $(A, +, .)$ and of a vector space $(A, +)$ over F such that the set A, the addition and the zero are the same in the ring and in the vector space and*

$$a(xy) = (ax)y = x(ay), \ a \in F, \ x, y \in A.$$

- *A commutative algebra is an algebra A with*

$$xy = yx, \ x, y \in A.$$

- *A normed algebra $(A, ||.||)$ over a field F is an algebra A with a norm $||.||$ that satisfies:*

  1. $||x|| \geq 0$, $x \in A$ and $||x|| = 0 \Leftrightarrow x = 0$,

  2. $||ax|| = |a|.||x||$, $x \in A$, $a \in F$,

  3. $||x + y|| \leq ||x|| + ||y||$, $x, y \in A$,

  4. $||xy|| \leq ||x||.||y||$.

- *A Banach algebra is a normed algebra $(A, \|.\|)$ that is also a Banach space for its norm, i.e. that is complete for the induced metric.*

- *Assume that A is a commutative Banach algebra and let $x \in A$. We define*

$$e_n(x) \triangleq \sum_{i=0}^{n} \frac{x^i}{i!}.$$

The sequence (e_i) is a Cauchy sequence in the Banach algebra A and thus it converges. We define the exponential of $x \in A$ by the limit

$$e^x = \lim_{n \to \infty} \sum_{i=0}^{n} \frac{x^i}{i!}.$$

Note that, since $e^x e^{-x} = 1$, the exponential of an element of a Banach algebra is always invertible.

- *An element $y \in A$ is said to have a logarithm if there exists $x \in A$ such that $y = e^x$. Elements that have a logarithm are invertible but invertible elements do not always have a logarithm.*

Theorem A.6 *Assume that A is a Banach algebra and suppose that $y \in A$ is invertible. If $x \in A$ satisfies*

$$\|x - y\| < \frac{1}{\|y^{-1}\|}$$

then x is invertible. ■

Appendix B

Analytic functions

Good references on analytic function theory include Hille [60], Conway [28], Nehari [93] and Goluzin [56].

Definition B.1 (Analytic functions) • *A domain (or, a region) is an open connected subset of* **C**.

  • *A complex valued function $f(z)$ is analytic[1] on Ω if it has a derivative at each point of Ω.*

We are mostly interested by functions that are analytic in a domain Ω and that are continuous on the boundary $\partial\Omega$. A typical example is given by the disc algebra.

Definition B.2 (Disc algebra) *The set of functions that are analytic in D and continuous on $\overline{D}$ is called the disc algebra and is denoted by $A(\overline{D})$.*

---

[1] Different authors, different countries, different terminologies. The possible choices for designating analytic functions is vast: monogenic, regular, regular-analytic, synectic (slightly old-fashioned, see Hille [60]) or holomorphic (the French trend, see Nehari p.59 [93]). All different names for the same thing: complex valued functions that have a derivative.

The next theorems are valid for functions that are analytic on a domain Ω and continuous on $\partial\Omega$.

Theorem B.1 (Maximum modulus) *Assume that $f(z)$ is analytic in a domain Ω and continuous on its boundary $\partial\Omega$. Then $|f(z)|$ attains its maximum on $\partial\Omega$.*

Theorem B.2 (Rouché's theorem) *Assume that $f(z)$ and $g(z)$ are analytic in a domain Ω and continuous on its boundary $\partial\Omega$. If $|f(z)| > |g(z)|$ for all $z \in \partial\Omega$ then $f(z)$ and $f(z) + g(z)$ have the same number of zeros in Ω.*

Theorem B.3 (Mergelyan's theorem) *Assume that K is a compact subset of $\mathbf{C}$, K^c is connected, $f(z)$ is continuous on K and analytic in the interior of K. Then associated to any strictly positive ϵ is a polynomial $p(z) \in \mathbf{R}[z]$ such that,*

$$|p(z) - f(z)| < \epsilon, \quad \text{for all } z \in K.$$

Theorem B.4 (Disc algebra) *The disc algebra is a Banach algebra.*

Definition B.3 (Normal families) *Let $G = (f_\alpha(z))$ be a family of functions that are analytic in a domain Ω. G is a normal family if from any sequence of functions in G it is possible to extract a subsequence which converges uniformly on any compact subset of Ω.*

As is customary, we extend the notion of convergence to that of convergence to the constant function ∞. We say that a sequence of functions $(f_i(z))$ converges uniformly to ∞ in any closed subset of Ω if for every $M > 0$ we have

$$|f_i(z)| > M$$

for i large enough and for all z in Ω.

The next two conditions are sufficient for a family G to form a normal family.

Theorem B.5 (Montel's theorem) *If any one of the next two conditions is satisfied, then the family G is a normal family.*

1. *G is locally uniformly bounded. That is*
   $\forall z_0 \in D,\ \exists M > 0,\ \exists r > 0$ such that $|f(z)| < M,\ \forall |z - z_0| < r,\ \forall f(z) \in G$.

2. *Each member of G has a lacunary value $a \in \mathbb{C}$ and a value $b \in \mathbb{C}$ that is attained at most p times. That is, there exists $p \in \mathbb{N}$ and $a, b \in \mathbb{C}$, $a \neq b$ such that:*

   (a) $\forall f(z) \in G,\ f(z) \neq a$ for $z \in \Omega$,

   (b) $\forall f(z) \in G$, the cardinality of the set defined by $\{z \in \Omega : f(z) = b\}$ is less than p.

Appendix C

Range of analytic functions

The references for this appendix are Ahlfors [1], Conway [28], Nehari [93], Hille [60] and Goluzin [56]. In each of these books we have picked results on the range of analytic functions.

Assume that $f(z)$ is a complex valued function that is analytic on the open unit disc D. The *range of $f(z)$ on D* is defined by $f(D) = \{f(z) : z \in D\}$. The value $a \in \mathbf{C}$ is a *lacunary value* of $f(z)$ on D if $a \notin f(D)$.

Our objective in this appendix is to describe characteristics of $f(D)$ that are as precise as possible and yet valid for classes of analytic functions that contains as many members as possible. We present two kinds of such results. The *first* ones concern functions that are normalized by the condition $f'(0) = 1$ and are in the form of filled ball results. The *second* family of results are on analytic functions $f(z)$ that have two lacunary values i.e. functions that have two complex values that are outside their range. These results are in the form of *empty ball* results.

C.1 Filled ball results

Suppose that, following the ideas developed in the introduction, we wish to find a family $\mathcal{F}$ of analytic functions that contains as many members as possible and such that any of its members contains a ball of fixed radius in its range.

Obviously, we can not tolerate to admit the constant functions $f(z) = a$ $(a \in \mathbb{C})$ in the family $\mathcal{F}$ since the range of such functions reduces to a single value. In order to exclude these cases we consider the family $\mathcal{F}$ of analytic functions that are normalized by the condition $f'(0) = 1$ and we refer to these functions as *normalized functions* (these correspond to the functions that have a convergent power series of the form $f(z) = a_0 + z + a_2 z^2 + a_3 z^3 + ...$).

Theorem C.1 *Bloch's theorem: Hille p.386 [60], Nehari p.363 [93].*
There exists a strictly positive constant b such that any analytic function on the open unit disc D that is normalized by $f'(0) = 1$ is such that there exists a disc $S \subset D$ with:

1. *$f(z)$ is bijective on S,*

2. *the range of $f(z)$ on S contains a ball of radius b.*

The largest such b over all normalized functions is denoted by B and is called Bloch's constant[1]. ∎

Proving that the non zero constant B exists and computing it explicitly are two rather different problems. We quote one of Bloch's

---

[1] Note, of course, that this constant is independent of the particular function $f(z)$.

original comments –in 1926– on the value of B (the two next quotations are taken from [12], 1926):

On pourrait, à l'aide de la démonstration, trouver des valeurs de ces constantes; mais seule présenterait de l'intérêt la détermination des valeurs exactes.

And, further in the same text:

Tous ces théorèmes pourront être précisés par diverses formules.

It is amusing to note that, more than sixty years after these comments were made, the value of Bloch's constant is still unknown[2]. Yet, the best bounds obtained are

$$0.433... \leq B \leq 0.472....$$

It was conjectured by Ahlfors in 1937 that the exact value of B is given by the upper bound above

$$B = \frac{\Gamma(\frac{1}{3})\Gamma(\frac{11}{12})}{(1 + \sqrt{3})^{\frac{1}{2}}\Gamma(\frac{1}{4})}.$$

This conjecture is still standing.

Bloch's theorem says that, given a normalized analytic function $f(z)$, it is possible to find a disc $S \subset D$ on which $f(z)$ is bijective and such that $f(S)$ contains a ball of radius B. From this it is clear that any normalized analytic function has a ball of radius B in its range on D. If we do not require the bijectivity condition the value B is not the best possible and this justifies the next theorem.

---

[2]The determination of Bloch's constant is proposed under the entry number 20 among 30 other problems of the same kind in J. Littlewood, *Some problems in real and complex analysis*, Heath mathematical monographs, 1968.

Theorem C.2 *Landau's theorem: Hille p.386 [60], Nehari p.363 [93].*

There exists a strictly positive constant l such that any analytic function on the open unit disc D that is normalized by $f'(0) = 1$ is such that the range of $f(z)$ on D contains a ball of radius l.

The largest such l is denoted by L and is called Landau's constant.

■

It is of course true that

$$B \leq L$$

and thus, since $0.43... \leq B$, we have $0.43... \leq L$.

The function $f(z) = z$ shows also that $L \leq 1$ and thus

$$0.43... \leq L \leq 1.$$

Sharper computations lead to

$$0.5 \leq L \leq 0.544....$$

But the exact computation of L has, so far, been out of reach. As in the case of Bloch's constant, Landau's constant is unknown.

We move now to a slightly different question. When an analytic function is normalized, it contains some ball of radius L in its range. What we do not know, however, is where the center of the ball lies. Imagine that we wish to fit a ball that has its center at the origin in the range of an analytic function. A natural way to expect this to be the case is to impose not only $f'(0) = 1$ but also $f(0) = 0$, which ensures that the origin is contained in the range.

An analytic function satisfying both at $f'(0) = 1$ and at $f(0) = 0$ is refered to as *doubly normalized* (these functions correspond to those that have a convergent power series of the form $f(z) = z + a_2 z^2 + ...)$. The question is then

> *does there exist a strictly positive constant k such that any doubly normalized analytic function has a ball of radius k centered at the origin in its range?*

The answer to this question is negative and this can be seen from the following example (the example is taken from Goluzin [56]).

Define $f_n(z) \triangleq \frac{1-(1-z)^n}{n}$. For any n the function $f_n(z)$ is

1. analytic in D,

2. such that $f(0) = 0$,

3. such that $f'(0) = 1$.

But also $f_n(z) = 1 \Leftrightarrow z = \frac{1}{n}$ and thus the range of the function $f_n(z)$ on D does not contain the ball centered at the origin and of radius $\frac{1}{n}$.

This example clearly shows that it is not always possible to find a ball centered at the origin and of fixed radius in the range of a doubly normalized analytic function. There exists, however, a segment of line that satisfies such a property.

Theorem C.3 *Bermant's theorem: Goluzin p.89 [56] .*
The range of any doubly normalized analytic function on D completely covers some segment of arbitary prenamed slope that contains the origin and is of length no less than $\frac{8\pi^2}{\Gamma^4(\frac{1}{4})} = 0.456946...$ This last constant is sharp. ∎

An immediate consequence of this is:

Theorem C.4 *Nehari p.328 [93].*
Assume that $f(z)$ is a normalized analytic function such that

$$f(z) = -f(-z), \ z \in D.$$

Then the image of the disk D by $f(z)$ completely covers some ball that contains the origin and is of radius no less than $\frac{4\pi^2}{\Gamma^4(\frac{1}{4})}$. ∎

Let us briefly recapitulate. The range on D of a normalized analytic function always contains a ball of radius L in its range. If $f(0) = 0$ then the range of $f(z)$ on D does not always contains a ball centered at the origin but it does contains a *segment* of line that includes the origin and is of length no less than a fixed value. If $f(z) = -f(-z)$ then $f(D)$ contains a ball centered at the origin.

There exist other constraints that ensure the existence of a ball around the origin that is entirely contained in the range. The next two theorems are examples of such results.

Theorem C.5 *Goluzin's theorem: Goluzin p.73 [56].*
Associated to each integer n that is greater than or equal to one is a positive constant ρ_n such that any doubly normalized analytic function $f(z)$ that has less than n zeros in D has a range on D containing a ball centered at the origin and of radius no less than ρ_n. ∎

An explicit value for ρ_n is known when $n = 1$: $\rho_1 = \frac{1}{16}$. This value is also the best possible for $n = 1$. The theorem for $n = 1$ has been nicknamed the $\frac{1}{16}$-*Theorem*[3] and it reads.

---

[3]This by analogy with the well-known *Koebe covering lemma*, or $\frac{1}{4}$-Theorem,

Theorem C.6 $\frac{1}{16}$-*theorem: Goluzin p.89 [56], Nehari p.323 [93].*
Any analytic
function on D that satisfies

    1. $f(z) = 0$, $z \in D \Leftrightarrow z = 0$,

    2. $f'(0) = 1$,

*contains an open ball of radius $\frac{1}{16}$ and of center $z = 0$ in its range
on D but not always a larger ball.* ∎

Another result of the same cask is:

Theorem C.7 *Goluzin p.75 [56].*
*Assume that $f(z)$ is a doubly normalized analytic function that
never takes the value $a \in \mathbb{C}$ for $z \in D$. Then there exists a positive
constant ρ such that the disk $|z| \leq \rho$ is completely covered by the
image of D under $f(z)$.* ∎

All these results are on the existence of filled balls in the range of
analytic functions. As is explained in the introduction, the next
section provides similar results on the existence of 'empty' balls.

C.2 Empty ball results

Analytic functions with two lacunary values have been extensively
studied in the past (see for example Bloch [12]). We give below
what we believe to be the two most important results in this field:

---

on univalent functions: an univalent function that satisfies $f(0) = 0$ and $f'(0) =$
1 always has a ball of radius $\frac{1}{4}$ in its range on D but not always a larger ball.
The same remains correct for p-valent functions. See Goluzin [56] for this.

Schottky's theorem (also called Picard-Schottky's theorem) and a theorem of Landau.

Theorem C.8 *Schottky's theorem: Ahlfors p.19 [1] or Hayman [59].*

Assume

that $f(z)$ is an analytic function on D that has 0 and 1 as lacunary values. Then [Ahlfors]:

$$|f(z)| \leq e^{(7+\max(0,\log|f(0)|))\frac{1+|z|}{1-|z|}}, \ z \in D.$$

A rewriting of this with sharper constants gives [Hayman]:

$$|f(z)| \leq (\max\{1,|f(0)|\})e^{\pi(\frac{1+|z|}{1-|z|})}, \ z \in D.$$

∎

If a function has 0 and 1 as lacunary values then its inverse is analytic and shares the same properties. Applying the above theorem to the inverse of $f(z)$ we get that under the same assumptions

$$|f(z)| \geq (\min\{1,|f(0)|\})e^{-\pi(\frac{1+|z|}{1-|z|})}, \ z \in D.$$

This last formulation clearly shows the empty ball formulation. Any analytic function that has 0 and 1 as lacunary values is bounded *below* on any compact subset of D by a value that depends only on the compact subset and on the value of the function at the origin.

The next result gives a tight relation between the value of $f(0)$ and that of $f'(0)$ for a function that has both 0 and 1 as lacunary values.

Theorem C.9 *Landau's theorem: Hempel [62].*

Assume that $f(z)$ is an analytic function on D that has 0 and 1 as lacunary values. Then

$$|f'(0)| \leq 2|f(0)| \left(|\log|f(0)|| + \frac{\Gamma^4(\frac{1}{4})}{4\pi^2} \right).$$

The value $\frac{\Gamma^4(\frac{1}{4})}{4\pi^2}$ is the best possible and is approximatively equal to 4.377. ■

In this theorem the constant is sharp in the sense that the theorem would not be true for a smaller constant.

The weird constants obtained in these theorems are introduced by means of the elliptic modular function.

To conclude, it is perhaps worth saying that the theorems of this appendix have no counterparts for real functions but are really specific to complex analytic functions.

Bibliography

[1] Ahlfors L., Conformal invariants, *Mc Graw Hill Series in higher mathematics*, 1973.

[2] Ackermann J., Uncertainty and control, Lecture Notes in Control and Information Sciences 70, *Springer-Verlag*, Berlin, 1985.

[3] Alos A., Stabilization of a class of plants with possible loss of outputs or actuator failures, *IEEE Trans. Automat. Control*, vol. 28, pp. 231-233, 1983.

[4] Anderson B., A note on the Youla-Bongiorno-Lu condition, *Automatica*, vol. 2, pp. 387-388, 1976.

[5] Anderson B., Bose N. and Jury E., Output feedback stabilization and related problems - solutions via decision methods, *IEEE Trans. Automat. Control*, vol. 20, pp. 387-388, 1975.

[6] Anderson B., Dasgupta S., Khargonekar P., Krause F. and Mansour M., Robust strict positive realness: characterization and construction, *IEEE Trans. Circuits and Systems*, vol. 37, pp. 869-876, 1990.

[7] Antoulas A. (ed.), Mathematical System Theory, *Springer-Verlag*, Berlin, 1991.

[8] Apostol T., Mathematical Analysis, *Addison-Wesley*, 1974.

[9] Atiyah M. and MacDonald I., Commutative Algebra, *Addison-Wesley*, 1969.

[10] Baker A., Transcendental number theory, *Cambridge University Press*, Cambridge, 1979.

[11] Bartlett A., Hollot C. and Lin H., Root location of an entire polytope of polynomials: it suffices to check the edges, *Math. Control Signals Systems*, vol. 1, pp. 61-71, 1988.

[12] Bloch A., Les fonctions holomorphes et méromorphes dans le cercle unité, *Mémorial des Sciences Mathématiques*, Académie des Sciences de Paris, Fascicule XX, 1926.

[13] Blondel V., A problem from control theory: solvability of equations over an Euclidean domain, *MSc dissertation in Pure Mathematics*, Imperial College, London, 1990.

[14] Blondel V., A counterexample to a simultaneous stabilization condition for systems with identical unstable poles and zeros, *Systems Control Lett.*, vol. 17, pp. 339-341, 1991.

[15] Blondel V., Mortini R. and Rupp R., Simultaneous stabilization in the disc algebra, technical report AP90.34, University of Louvain, 1990.

[16] Blondel V., Gevers M., Mortini R. and Rupp R., Simultaneous stabilization of three or more plants: conditions on the real axis do not suffice, to appear in *SIAM J. Control and Optimiz.*, 1993.

[17] Blondel V., Campion G. and Gevers M., A sufficient condition for simultaneous stabilization, to appear in *IEEE Trans. Automat. Control*, 1993.

[18] Blondel V., Campion G. and Gevers M., Avoidance and intersection in the complex plane: a tool for simultaneous stabilization, *Proc. IEEE 30th Conf. on Decision and Control*, Brighton, UK, vol. 1, pp. 285-290, 1991.

[19] Blondel V., Gevers M., Mortini R. and Rupp R., Stabilizable by a stable and by an inverse stable but not by a stable and inverse stable, *Proc. IEEE 31st Conf. on Decision and Control*, Tucson, USA, 1992.

[20] Blondel V., Stabilization with respect to a general domain of stability, *Proc. 2nd Int. Sympos. Implicit Robust Systems*, Warsaw, Poland, pp. 33-37, 1991.

[21] Blondel V., Simultaneous stabilization of linear systems: mathematical solutions, related problems and equivalent formulations, *PhD Thesis*, Catholic University of Louvain, Louvain-la-Neuve, Belgium, 1992.

[22] Blondel V. and Gevers M., The simultaneous stabilizability question of three linear systems is rationally undecidable, to appear in *J. of Math. Control, Signal, and Systems*, 1993.

[23] Brockett R., Some geometric questions in the theory of linear systems, *IEEE Trans. Automat. Control*, vol. 21, pp. 449-455, 1976.

[24] Callier F. and Desoer C., Multivariable Feedback Systems, *Springer-Verlag*, New York, 1982.

[25] Callier F. and Desoer C., Stabilization, tracking and disturbance rejection in multivariable convolution systems, *Annales
de la Société Scientifique de Bruxelles*, vol. 94, pp. 7-51, 1980.

[26] Cavallo A., De Maria G., A polynomial approach to simultaneous stabilization, *Proc. 30th IEEE Conf. on Decision and
Control*, Brighton, UK, 1991.

[27] Cohn P., Algebra, John Wiley, vol. I and II, 1974.

[28] Conway J., Functions of one complex variable, *Springer-
Verlag*, 1973.

[29] Corach G. and Suarez F., Stable rank in holomorphic functions algebras, *Illinois J. Math.*, vol. 29, pp. 627-639, 1985.

[30] Dasgupta S., Kharitonov's theorem revisited, *Systems Control Lett.*, vol. 11, pp. 381-384, 1988.

[31] Dasgupta S. and Bhagwat A., Conditions for designing
strictly positive real transfer functions for adaptive output
error identification, *IEEE Trans. Automat. Control*, vol. 34,
pp. 731-736, 1987.

[32] Delsarte P., Quelques observations à propos du papier "simultaneous stabilization of three or more plants", *Personal
communication*, January 1992.

[33] Desoer C., Liu R., Murray J., Saeks R., Feedback system
design: the fractional representation approach to analysis and
synthesis, *IEEE Trans. Automat. Control*, vol. 28, pp. 399-
412, 1980.

[34] Debowski A. and Kurylowicz A., Simultaneous stabilization of linear single-input/single output plants, *Int. J. Control*, vol. 44, pp. 1257-1264, 1986.

[35] Djaferis T., To stabilize a k real parameter affine family of plants it suffices to simultaneously stabilize 4^k polynomials, *Systems Control Lett.*, vol. 16, pp. 187-193, 1991.

[36] Dorato P., Park H. and Li Y., An algorithm for interpolation with units in H^∞, with applications to feedback stabilization, *Automatica*, vol. 25, pp. 427-430, 1989.

[37] Dorato P., Li Y. and Park H., U-parameter design: feedback system design with guaranteed robust stability, in Milanese M., Tempo R. and Vicino A. (eds), *Robustness in identification and control*, pp. 321-327, 1989.

[38] Duren P., The theory of H^p spaces, *Academic Press*, New York, 1970.

[39] Emre E., Simultaneous stabilization with fixed closed loop characteristic polynomial, *IEEE Trans. Automat. Control*, vol. 28, pp. 103-104, 1983.

[40] Emre E., On necessary and sufficient conditions for regulation of linear systems over rings, *SIAM J. Control and Optimiz.*, vol. 20, pp. 155-160, 1982.

[41] El-Sakkary A., Estimating robustness on the Riemann sphere, *Int. J. Control*, vol. 49, pp. 561-567, 1989.

[42] Francis B., Helton W. and Zames G., H_∞-optimal feedback controllers for linear multivariable systems, *IEEE Trans. Automat. Control*, vol. 29, pp. 888-900, 1984.

174

[43] Francis B. and Zames G., On H_∞-optimal sensitivity theory for siso feedback systems, *IEEE Trans. Automat. Control*, vol. 29, pp. 9-16, 1984.

[44] Francis B., A course in H_∞ control theory, *Springer-Verlag*, Berlin, 1987.

[45] Fu M., An introduction to the parametric approach to robust stability and robust control, lecture notes of seminars given at the University of Louvain, Belgium, 1990.

[46] Fu M., Dasgupta S. and Blondel V., Robust stability under a class of nonlinear parametric perturbations, *Proc. Amer. Control Conf.*, San Diego, USA, pp. 2542-2547, 1990.

[47] Gaier D., Lectures on complex approximation, *Birkhäuser*, Basel, 1987.

[48] Gantmacher F., Matrix theory, vol. I and II, *Chelsea*, New York, 1959.

[49] Gelfond A., The solution of equation with integer, *Golden gate books*, 1961.

[50] Ghosh B. and Byrnes C., Simultaneous stabilization and pole-placement by nonswitching dynamic compensation, *IEEE Trans. Automat. Control*, vol. 28, pp. 735-741, 1983.

[51] Ghosh B., Some new results on the simultaneous stabilizability of a family of single input single output systems, *Systems Control Lett.*, vol. 6, pp. 39-45, 1985.

[52] Ghosh B., Transcendental and interpolation methods in simultaneous stabilization and simultaneous partial pole place-

ment problems, *SIAM J. Control and Optimiz.*, vol. 24, pp. 1091-1109, 1986.

[53] Ghosh B., An approach to simultaneous system design. Part 1: Semialgebraic geometric methods, *SIAM J. Control and Optimiz.*, vol. 24, pp. 480-496, 1986.

[54] Ghosh B., An approach to simultaneous system design. Part 2, *SIAM J. Control and Optimiz.*, vol. 26, pp. 919-963, 1988.

[55] Ghosh B., Simultaneous partial pole placement: a new approach to multimode system design, *IEEE Trans. Automat. Control*, vol. 31, pp.440-443, 1986.

[56] Goluzin G., Geometric theory of functions of a complex variable, Translation of Math. Monographs, vol. 26, *American Math. Society*, 1969.

[57] Goodearl K. and Menal P., Stable range one for rings with many units, *J. Pure and Applied Algebra*, vol. 54, pp. 261-287, 1988.

[58] Gündes A. and Desoer C., Algebraic theory of linear feedback systems with full and decentralized compensators, Lecture Notes in control and information sciences, vol. 142, *Springer-Verlag*, 1990.

[59] Hayman, Some remarks on Schottky's theorem, *Math. Proc. Cambridge Philos. Soc.*, vol. 43, pp. 442-454, 1947.

[60] Hille E., Analytic function theory, *Ginn and Co.*, 2 vol., 1959.

[61] Helton J., Worst case analysis in the frequency domain, *IEEE Trans. Automat. Control*, vol. 30, pp. 1154-1170, 1985.

[62] Hempel J., The Poincaré metric on the twice punctured plane and the theorems of Landau and Schottky, *J. London Math. Soc.*, vol. 20, pp. 435-445, 1980.

[63] Henrici P., Applied and computational complex analysis, vol. 1, *Wiley-interscience*, New-York, 1974.

[64] Hoffman K., Banach spaces of analytic functions, *Prentice Hall*, 1962.

[65] Hollot C., Kharitonov-like results in the space of Markov parameters, *IEEE Trans. Automat. Control*, vol. 34, pp. 536-538, 1989.

[66] Horowitz I., Synthesis of feedback systems, *Academic Press*, 1963.

[67] Hung N. and Anderson B., Triangularization technique for the design of multivariable control systems, *IEEE Trans. Automat. Control*, vol. 24, pp. 455-460, 1979.

[68] Jacobson N., Lectures in abstract algebra, *Van Nostrand*, New York, 1953.

[69] Jenkins J., On explicit bounds in Landau's theorem. II, *Can. J. Math.*, vol. 33, pp. 559-562, 1981.

[70] Jensen C., Some curiosities of rings of analytic functions, *J. Pure and Applied Algebra*, vol. 38, pp. 277-283, 1985.

[71] Jones P., Marshall D. and Wolff T., Stable rank of the disc algebra, *Proc. Amer. Math. Soc.*, vol. 96, pp. 603-604, 1986.

[72] Kabamba P. and Yang C., Simultaneous controller design for linear time invariant systems, *IEEE Trans. Automat. Control*, vol. 36, pp. 106-111, 1991.

[73] Kailath T., Linear Systems, *Prentice-Hall*, 1980.

[74] Khargonekar P. and Tannenbaum A., Non-Euclidean metrics and the robust stabilization of systems with parametric uncertainty, *IEEE Trans. Automat. Control*, vol. 30, pp. 1005-1013, 1985.

[75] Khargonekar P. and Ozgüler B., The ring of stable rational functions: algebraic properties, *Proc. IEEE 21st Conf. on Decision and Control*, pp. 402-407, 1982.

[76] Khargonekar P. and Sontag E. , On the relation between stable matrix fraction factorizations and regulable realizations of linear systems over rings, *IEEE Trans. Automat. Control*, vol. 27, pp. 627-638, 1982.

[77] Kharitonov V., Asymptotic stability of an equilibrium position of a family of systems of linear differential equations, *Differetsial'nye Uravneniya*, vol. 14, pp. 2086-2088, 1978.

[78] Kimura H., Robust stabilizability for a class of transfer functions, *IEEE Trans. Automat. Control*, vol. 29, pp. 788-793, 1984.

[79] Kinnaert M. and Blondel V., Pole placement with a stable controller, *Automatica*, vol. 28, pp. 935-945, 1992.

[80] Kučera V., Discrete linear control: the polynomial equation approach, *Wiley*, New York, 1979.

[81] Kwakernaak H., A condition for robust stabilizability, *Systems Control Lett.*, vol. 2, pp. 1005-1013, 1985.

[82] Kwakernaak H., Minimax frequency domain performance and robustness optimization of linear feedback systems, *IEEE Trans. Automat. Control*, vol. 30, pp. 994-1004, 1985.

[83] Kwakernaak H. and Sivan R., Linear optimal control systems, *Wiley*, New-York, 1972.

[84] Lang S., Structures algébriques, *InterEditions*, Paris, 1967.

[85] Lunze J., Robust multivariable feedback control, *Prentice Hall*, New York, 1989.

[86] Limebeer D. and Anderson B., An interpolation theory approach to H_∞ controller degree bounds, *Linear Algebra and its Applications*, vol. 98, pp. 347-386, 1988.

[87] McDuffee, Theory of matrices, *Chelsea*, New York, 1946.

[88] McFarlane D. and Glover K., Robust controller design using normalized coprime factor plant descriptions, Lecture notes in control and information sciences, vol. 138, *Springer-Verlag*, 1990.

[89] Marden M., The geometry of the zeros of a polynomial in a complex variable, *Amer. Math. Soc.*, 1949.

[90] Maeda M. and Vidyasagar M., Some results on simultaneous stabilization, *Systems Control Lett.*, vol. 5, pp. 205-208, 1984.

[91] Minto K. and Vidyasagar M., A state-space approach to simultaneous stabilization, *Control Theory and Adv. Technol.*, vol. 2, pp. 39-64, 1986.

[92] Minda D., Inequalities for the hyperbolic metric and applications to geometric function theory, in *Lecture notes in Mathematics*, vol. 1275, pp. 235-252, 1984.

[93] Nehari Z., Conformal mapping, *Internat. Series in Pure and Applied Math.*, McGraw-Hill, 1952.

[94] Obinata G. and Moore J., Characterization of controllers in simultaneous stabilization, *Systems Control Lett.*, vol. 10, pp. 333-340, 1988.

[95] Ohta Y., Maeda H. and Kodama S., Unit interpolation in H_∞: bounds of norm and degree of interpolants, *Systems Control Lett.*, vol. 17, pp. 251-256, 1991.

[96] Rudin W., Real and complex analysis, *McGraw-Hill*, 1986.

[97] Pernebo L., Algebraic control theory for multivariable systems, *PhD Thesis*, Lund Institute of Technology, Sweden, 1978.

[98] Pólya G. and Szegö G., Problems and theorems in analysis, *Springer-Verlag*, 1972.

[99] Rupp R., Stable rank of subalgebras of the disc algebra, *Proc. Amer. Math. Soc.*, vol. 108, pp. 137-142, 1990.

[100] Rupp R., Über den Bass-stable-rank komplexer funktionenalgebren, *PhD Thesis*, Universität Karlsruhe, Germany, 1988.

[101] Rupp R., A covering theorem for a composite class of analytic functions, preprint, 1993.

[102] Saeks R. and Murray J., Fractional representation, algebraic geometry and the simultaneous stabilization problem, *IEEE Trans. Automat. Control*, vol. 27, pp. 895-903, 1982.

[103] Simmons G., Topology and modern analysis, *McGraw-Hill*, 1963.

[104] Smith M. and Sondergeld K., On the order of stable compensators, *Automatica*, vol. 22, pp. 127-129, 1986.

[105] Ueda H., On the zero-one-pole set of a meromorphic function, *Kodai Math. J.*, vol. 12, pp. 9-22, 1989.

[106] Vardulakis A., Limebeer D. and Karcanias N., Structure and Smith MacMillan form of a rational matrix at infinity, *Int. J. Control*, vol. 35, pp. 701-725, 1982.

[107] Vaserstein L., Bass's first stable range condition, *J. Pure Applied Algebra*, vol. 34, pp. 319-330, 1984.

[108] Vidyasagar M., Control System Synthesis: a factorization approach, *MIT Press*, 1985.

[109] Vidyasagar M., Some results on simultaneous stabilization with multiple domains of stability, *Automatica*, vol. 23, pp. 535-540, 1987.

[110] Vidyasagar M. and Viswanadham N., Algebraic design techniques for reliable stabilization, *IEEE Trans. Automat. Control*, vol. 27, pp. 1085-1095, 1982.

[111] Vidyasagar, M. Schneider H. and Francis B., Algebraic and topological aspects of feedback stabilization, *IEEE Trans. Automatic. Control*, vol. 27, pp. 880-894, 1982.

[112] Vidyasagar M., A state-space interpretation of simultaneous stabilization, *IEEE Trans. Automat. Control*, vol. 33, pp. 506-508, 1988.

[113] Vidyasagar M. and Viswanadham N., Reliable stabilization using a multi-controller configuration, *Automatica*, vol. 21, pp. 599-602, 1985.

[114] Vidyasagar M., Levy B. and Viswanadham N., A note on the generecity of simultaneous stabilizability and pole assignability, *Circuits Systems Signal Process.*, vol. 5, pp. 371-387, 1986.

[115] Walsh J., Interpolation and approximation by rational functions in the complex plane, *AMS Colloquium Publ.*, Providence, R.I., 1935.

[116] Wei K. and Barmish B., An iterative design procedure for simultaneous stabilization of MIMO systems, *Automatica*, vol. 24, pp. 643-652, 1988.

[117] Wei K., Simultaneous pole assignment for a class of linear time invariant siso systems, *Proc. IEEE 28th Conf. on Decision and Control*, Tampa, Florida, pp. 1247-1252, 1989.

[118] Wei K., Stabilization of a linear plant via a stable compensator having no real unstable zeros, *Systems and Control Lett.*, vol. 15, pp. 259-264, 1990.

[119] Wei K., The solution of a transcendental problem and its application in simultaneous stabilization problems, *DLR Technical Report*, reference R38-91, 1991.

[120] Wei K. and Yedavalli R., Robust stabilizability for linear systems with both parameter variation and unstructured uncertainty, *IEEE Trans. Automat. Control*, vol. 34, pp. 149-156, 1989.

[121] Weinmann A., Uncertain models and robust control, *Springer-Verlag*, Wien, 1991.

[122] Weyl H., Symmetry, *Princeton University Press*, Princeton, 1952.

[123] Youla D., Bongiorno J. and Lu C., Single-loop feedback stabilization of linear multivariable plants, *Automatica*, vol. 10, pp. 159-173, 1974.

[124] Youla D., Bongiorno J. and Jabr H., Modern Wiener-Hopf design of optimal controllers, part I: the single input case, *IEEE Trans. Automat. Control*, vol. 21, pp. 3-148, 1976.

[125] Youla D., Jabr H. and Bongiorno J, Modern Wiener-Hopf design of optimal controllers, part II: the multivariable case, *IEEE Trans. Automat. Control*, vol. 21, pp. 319-338, 1976.

[126] Zames G., Feedback and optimal sensitivity: model reference transformations, multiplicative seminorms and approximate inverses, *IEEE Trans. Automat. Control*, vol. 26, pp. 301-320, 1981.

Index

Algebraic number 135

Analytic function 157

Avoidance 30, 31

Banach algebra 156

Biproper 17

Bistable rational function 17

Cauchy index 89

Constant

    Bloch's constant 161

    Landau's constant 163

Coprime 24, 152

Disc algebra 157

Elementary operation 134

Field of fractions 153

Fractional factorization 24

Greatest common divisor 151

High frequency gain 126

Interlacing property

    parity interlacing property
    71, 73

    even interlacing property
    83

    3-interlacing property 79

    k-interlacing property 84

Intersection 30

    simultaneous intersection 30,
    75

Inverstable 17

Minimum phase 18

Normalized function 161

Plant 16

Pole 16

Proper 17

Principal ideal domain 152

Stabilization 20

    bistable stabilization 21, 38

    external stabilization 18

    internal stabilization 18

    strong stabilization 5, 21,
    38

    simultaneous stabilization
    xiii, 4, 21

    $R_{+\infty}$-stabilization 69

Strictly proper 17

System 16

Theorem of –

    Bermant 164

    Bloch 161

Goluzin 165

Landau 167

Mergeylan 158

Montel 159

Rouché 158

Schottky 167

the maximum modulus 158

$\frac{1}{16}$-theorem 166

Transcendental number 136

Unit 151

Winding number 90

Zero 16

$A(\overline{D})$ **xvi, 108**

$\mathbf{C}_\infty, \mathbf{C}_{+\infty}, \mathbf{R}_\infty, \mathbf{R}_{+\infty}$ **xv**

$D, \overline{D}, \partial D$ **xv, 40**

$\mathbf{R}(s), \mathbf{R}[s], \mathbf{R}'(s)$ **xvi**

$S, S(\Lambda)$ **xvi, 23, 41**

$Stab(p)$ 47

$Stab_f(p)$ 56

$Stab_p(p)$ 57

$U, U(\Lambda)$ **xvi, 23, 41**

Lecture Notes in Control and Information Sciences

Edited by M. Thoma

1989–1993 Published Titles:

Vol. 135: Nijmeijer, Hendrik; Schumacher, Johannes M. (Eds.)
Three Decades of Mathematical System Theory. A Collection of Surveys at the Occasion of the 50th Birthday of Jan C. Willems.
562 pp. 1989 [3-540-51605-0]

Vol. 136: Zabczyk, Jerzy W. (Ed.)
Stochastic Systems and Optimization.
Proceedings of the 6th IFIP WG 7.1 Working Conference, Warsaw, Poland, September 12-16, 1988.
374 pp. 1989 [3-540-51619-0]

Vol. 137: Shah, Sirish L.; Dumont, Guy (Eds.)
Adaptive Control Strategies for Industrial Use.
Proceedings of a Workshop held in Kananaskis, Canada, 1988.
360 pp. 1989 [3-540-51869-X]

Vol. 138: McFarlane, Duncan C.; Glover, Keith
Robust Controller Design Using Normalized Coprime Factor Plant Descriptions.
206 pp. 1990 [3-540-51851-7]

Vol. 139: Hayward, Vincent; Khatib, Oussama (Eds.)
Experimental Robotics I. The First International Symposium, Montreal, June 19-21, 1989.
613 pp. 1990 [3-540-52182-8]

Vol. 140: Gajic, Zoran; Petkovski, Djordjija; Shen, Xuemin (Eds.)
Singularly Perturbed and Weakly Coupled Linear Control Systems. A Recursive Approach.
202 pp. 1990 [3-540-52333-2]

Vol. 141: Gutman, Shaul
Root Clustering in Parameter Space.
153 pp. 1990 [3-540-52361-8]

Vol. 142: Gündes, A. Nazli; Desoer, Charles A.
Algebraic Theory of Linear Feedback Systems with Full and Decentralized Compensators.
176 pp. 1990 [3-540-52476-2]

Vol. 143: Sebastian, H.-J.; Tammer, K. (Eds.)
System Modelling and Optimizaton.
Proceedings of the 14th IFIP Conference, Leipzig, GDR, July 3-7, 1989.
960 pp. 1990 [3-540-52659-5]

Vol. 144: Bensoussan, A.; Lions, J.L. (Eds.)
Analysis and Optimization of Systems.
Proceedings of the 9th International Conference. Antibes, June 12-15, 1990.
992 pp. 1990 [3-540-52630-7]

Vol. 145: Subrahmanyam, M. Bala
Optimal Control with a Worst-Case Performance Criterion and Applications.
133 pp. 1990 [3-540-52822-9]

Vol. 146: Mustafa, Denis; Glover, Keith
Minimum Entropy H Control.
144 pp. 1990 [3-540-52947-0]

Vol. 147: Zolesio, J.P. (Ed.)
Stabilization of Flexible Structures. Third Working Conference, Montpellier, France, January 1989.
327 pp. 1991 [3-540-53161-0]

Vol. 148: Not published

Vol. 149: Hoffmann, Karl H; Krabs, Werner (Eds.)
Optimal Control of Partial Differential Equations. Proceedings of IFIP WG 7.2 - International Conference. Irsee, April, 9-12, 1990.
245 pp. 1991 [3-540-53591-8]

Vol. 150: Habets, Luc C.
Robust Stabilization in the Gap-topology.
126 pp. 1991 [3-540-53466-0]

Vol. 151: Skowronski, J.M.; Flashner, H.; Guttalu, R.S. (Eds.)
Mechanics and Control. Proceedings of the 3rd Workshop on Control Mechanics, in Honor of the 65th Birthday of George Leitmann, January 22-24, 1990, University of Southern California.
497 pp. 1991 [3-540-53517-9]

Vol. 152: Aplevich, J. Dwight
Implicit Linear Systems.
176 pp. 1991 [3-540-53537-3]

Vol. 153: Hajek, Otomar
Control Theory in the Plane.
269 pp. 1991 [3-540-53553-5]

Vol. 154: Kurzhanski, Alexander; Laseicka, Irena (Eds.)
Modelling and Inverse Problems of Control for Distributed Parameter Systems. Proceedings of IFIP WG 7.2 - IIASA Conference, Laxenburg, Austria, July 1989.
170 pp. 1991 [3-540-53583-7]

Vol. 155: Bouvet, Michel; Bienvenu, Georges (Eds.)
High-Resolution Methods in Underwater Acoustics.
244 pp. 1991 [3-540-53716-3]

Vol. 156: Hämäläinen, Raimo P.; Ehtamo, Harri K. (Eds.)
Differential Games - Developments in Modelling and Computation. Proceedings of the Fourth International Symposium on Differential Games and Applications, August 9-10, 1990, Helsinki University of Technology, Finland.
292 pp. 1991 [3-540-53787-2]

Vol. 157: Hämäläinen, Raimo P.; Ehtamo, Harri K. (Eds.)
Dynamic Games in Economic Analysis. Proceedings of the Fourth International Symposium on Differential Games and Applications. August 9-10, 1990, Helsinki University of Technology, Finland.
311 pp. 1991 [3-540-53785-6]

Vol. 158: Warwick, Kevin; Karny, Miroslav; Halouskova, Alena (Eds.)
Advanced Methods in Adaptive Control for Industrial Applications.
331 pp. 1991 [3-540-53835-6]

Vol. 159: Li, Xunjing; Yong, Jiongmin (Eds.)
Control Theory of Distributed Parameter Systems and Applications. Proceedings of the IFIP WG 7.2 Working Conference, Shanghai, China, May 6-9, 1990.
219 pp. 1991 [3-540-53894-1]

Vol. 160: Kokotovic, Petar V. (Ed.)
Foundations of Adaptive Control.
525 pp. 1991 [3-540-54020-2]

Vol. 161: Gerencser, L.; Caines, P.E. (Eds.)
Topics in Stochastic Systems: Modelling, Estimation and Adaptive Control.
1991 [3-540-54133-0]

Vol. 162: Canudas de Wit, C. (Ed.)
Advanced Robot Control. Proceedings of the International Workshop on Nonlinear and Adaptive Control: Issues in Robotics, Grenoble, France, November 21-23, 1990.
Approx. 330 pp. 1991 [3-540-54169-1]

Vol. 163: Mehrmann, Volker L.
The Autonomous Linear Quadratic Control Problem. Theory and Numerical Solution.
177 pp. 1991 [3-540-54170-5]

Vol. 164: Lasiecka, Irena; Triggiani, Roberto
Differential and Algebraic Riccati Equations with Application to Boundary/Point Control Problems: Continuous Theory and Approximation Theory.
160 pp. 1991 [3-540-54339-2]

Vol. 165: Jacob, Gerard; Lamnabhi-Lagarrigue, F. (Eds.)
Algebraic Computing in Control. Proceedings of the First European Conference, Paris, March 13-15, 1991.
384 pp. 1991 [3-540-54408-9]

Vol. 166: Wegen, Leonardus L. van der
Local Disturbance Decoupling with Stability for Nonlinear Systems.
135 pp. 1991 [3-540-54543-3]

Vol. 167: Rao, Ming
Integrated System for Intelligent Control.
133 pp. 1992 [3-540-54913-7]

Vol. 168: Dorato, Peter; Fortuna, Luigi;
Muscato, Giovanni
Robust Control for Unstructured Perturbations:
An Introduction.
118 pp. 1992 [3-540-54920-X]

Vol. 169: Kuntzevich, Vsevolod M.; Lychak,
Michael
Guaranteed Estimates, Adaptation and
Robustness in Control Systems.
209 pp. 1992 [3-540-54925-0]

Vol. 170: Skowronski, Janislaw M.; Flashner,
Henryk; Guttalu, Ramesh S. (Eds.)
Mechanics and Control. Proceedings of the 4th
Workshop on Control Mechanics, January
21-23, 1991, University of Southern
California, USA.
302 pp. 1992 [3-540-54954-4]

Vol. 171: Stefanidis, P.; Paplinski, A.P.;
Gibbard, M.J.
Numerical Operations with Polynomial
Matrices: Application to Multi-Variable
Dynamic Compensator Design.
206 pp. 1992 [3-540-54992-7]

Vol. 172: Tolle, H.; Ersü, E.
Neurocontrol: Learning Control Systems
Inspired by Neuronal Architectures and Human
Problem Solving Strategies.
220 pp. 1992 [3-540-55057-7]

Vol. 173: Krabs, W.
On Moment Theory and Controllability of
Non-Dimensional Vibrating Systems and
Heating Processes.
174 pp. 1992 [3-540-55102-6]

Vol. 174: Beulens, A.J. (Ed.)
Optimization-Based Computer-Aided Modelling
and Design. Proceedings of the First Working
Conference of the New IFIP TC 7.6 Working
Group, The Hague, The Netherlands, 1991.
268 pp. 1992 [3-540-55135-2]

Vol. 175: Rogers, E.T.A.; Owens, D.H.
Stability Analysis for Linear Repetitive
Processes.
197 pp. 1992 [3-540-55264-2]

Vol. 176: Rozovskii, B.L.; Sowers, R.B. (Eds.)
Stochastic Partial Differential Equations and
their Applications. Proceedings of IFIP WG 7.1
International Conference, June 6-8, 1991,
University of North Carolina at Charlotte, USA.
251 pp. 1992 [3-540-55292-8]

Vol. 177: Karatzas, I.; Ocone, D. (Eds.)
Applied Stochastic Analysis. Proceedings of a
US-French Workshop, Rutgers University, New
Brunswick, N.J., April 29-May 2, 1991.
317 pp. 1992 [3-540-55296-0]

Vol. 178: Zolésio, J.P. (Ed.)
Boundary Control and Boundary Variation.
Proceedings of IFIP WG 7.2 Conference,
Sophia- Antipolis,France, October 15-17,
1990.
392 pp. 1992 [3-540-55351-7]

Vol. 179: Jiang, Z.H.; Schaufelberger, W.
Block Pulse Functions and Their Applications in
Control Systems.
237 pp. 1992 [3-540-55369-X]

Vol. 180: Kall, P. (Ed.)
System Modelling and Optimization.
Proceedings of the 15th IFIP Conference,
Zurich, Switzerland, September 2-6, 1991.
969 pp. 1992 [3-540-55577-3]

Vol. 181: Drane, C.R.
Positioning Systems - A Unified Approach.
168 pp. 1992 [3-540-55850-0]

Vol. 182: Hagenauer, J. (Ed.)
Advanced Methods for Satellite and Deep
Space Communications. Proceedings of an
International Seminar Organized by Deutsche
Forschungsanstalt für Luft-und Raumfahrt
(DLR), Bonn, Germany, September 1992.
196 pp. 1992 [3-540-55851-9]

Vol. 183: Hosoe, S. (Ed.)
Robust Control. Proceesings of a Workshop
held in Tokyo, Japan, June 23-24, 1991.
225 pp. 1992 [3-540-55961-2]

Vol. 184: Duncan, T.E.; Pasik-Duncan, B.
(Eds.)
Stochastic Theory and Adaptive Control.
Proceedings of a Workshop held in Lawrence,
Kansas, September 26-28, 1991.
500 pages. 1992 [3-540-55962-0]

Vol. 185: Curtain, R.F. (Ed.); Bensoussan, A.; Lions, J.L.(Honorary Eds.)
Analysis and Optimization of Systems: State and Frequency Domain Approaches for Infinite-Dimensional Systems. Proceedings of the 10th · International Conference, Sophia-Antipolis, France, June 9-12, 1992.
648 pp. 1993 [3-540-56155-2]

Vol. 186: Sreenath, N.
Systems Representation of Global Climate Change Models. Foundation for a Systems Science Approach.
288 pp. 1993 [3-540-19824-5]

Vol. 187: Morecki, A.; Bianchi, G.; Jaworeck, K. (Eds.)
RoManSy 9: Proceedings of the Ninth CISM-IFToMM Symposium on Theory and Practice of Robots and Manipulators.
476 pp. 1993 [3-540-19834-2]

Vol. 188: Naidu, D. Subbaram
Aeroassisted Orbital Transfer: Guidance and Control Strategies.
192 pp. 1993 [3-540-19819-9]

Vol. 189: Ilchmann, Achim
Non-Identifier-Based High-Gain Adaptive Control
220 pp. 1993 [3-540-19845-8]

Vol. 190: Chatila, R; Hirzinger, G (Eds.)
Experimental Robotics II: The 2nd International Symposium, Toulouse, France, June 25-27 1991.
576 pp (approx.) 1993 [3-540-19851-2]